THE STRUCTURE AND COMPOSITION OF ANIMAL PRODUCTS

THE STRUCTURE AND COMPOSITION OF ANIMAL PRODUCTS

Giriraj Prajapati

RANDOM PUBLICATIONS
NEW DELHI - 110 002 (INDIA)

The Structure and Composition of Animal Products

ISBN 978-93-51116-90-5

Published in 2015 in India by

RANDOM PUBLICATIONS

4376-A/4B, Gali Murari Lal, Ansari Road
New Delhi-110 002
Phone: +9111-43580356, 23289044
E-mail: randomexports@gmail.com; sales@randompublications.com; info@randompublications.com

Reprinted 2018

Type Setting by: Friends Media, Delhi-110089
Digitally Printed at: Replika Press Pvt. Ltd.

Preface

Animals play very important roles in our lives. Animals are pets, they are raised as food and they provide products important to everyday life. You may not realize how many things come from animals. An animal product is any material derived from the body of an animal. Examples are fat, flesh, blood, milk, eggs, and lesser known products, such as isinglass and rennet. Animal by-products are carcasses and parts of carcasses from slaughterhouses, animal shelters, zoos and veterinarians, and products of animal origin not intended for human consumption, including catering waste. These products may go through a process known as "rendering" to be made into human and non-human foodstuffs, fats, and other material that can be sold to make commercial products such as cosmetics, paint, cleaners, polishes, glue, soap and ink. The sale of animal by-products allows the meat industry to compete economically with industries selling sources of vegetable protein. Generally, products made from fossilized or decomposed animals, such as petroleum formed from the ancient remains of marine animals, are not considered animal products. Crops grown in soil fertilized with animal remains are rarely characterized as animal products. Apart from breastfed infants, the human consumption of dairy products is sourced primarily from the milk of cows, water buffaloes, goats, sheep, yaks, horses, camels, domestic buffaloes, and other mammals. Dairy products are commonly found in European, Middle Eastern, and Indian cuisine, whereas aside from Mongolian cuisine they are little-known in traditional East Asian cuisine.

Leather is a durable and flexible material created by the tanning of animal rawhide and skin, often cattle hide. It can be produced through manufacturing processes ranging from cottage industry to heavy industry. Wool is the textile fibre obtained from sheep and certain other animals, including cashmere from goats, mohair from goats, qiviut from muskoxen, angora from rabbits, and other types of wool from camelids. Animal Products plays an important role in the socio-economic life of India. It is a rich source of high quality of animal products such as milk, meat and eggs. India has emerged as the largest producer of milk with 16.43 percent share in total milk production in the world. India accounts for about 4.95 percent of the global egg production and also the largest population of milch animals in the world, with 112.9 million buffaloes, 157 million goat and 74.5 million sheep. Exports of animal products represent an important and significant contribution to the Indian Agriculture sector. The export of Animal Products includes Buffalo meat, Sheep/Goat meat, Poultry products, Animal Casings, Milk and Milk products and Honey etc.

This book is intended as a standard reference book for dairy and animal science courses.

I thank all members of my team who have helped in the preparation of the book. My special thanks go to "Random Publications" who have published the book.

— Giriraj Prajapati

Contents

Chapter 1

Reproduction in Mammals

Production of new individuals along a leaf margin of the air plant, *Kalanchoë pinnata*. The small plant in front is about 1 cm tall. The concept of "individual" is obviously stretched by this asexual reproductive process. Reproduction is the biological process by which new individual organisms are produced. Reproduction is a fundamental feature of all known life; each individual organism exists as the result of reproduction. The known methods of reproduction are broadly grouped into two main types: sexual and asexual. In asexual reproduction, an individual can reproduce without involvement with another individual of that species. The division of a bacterial cell into two daughter cells is an example of asexual reproduction. Asexual reproduction is not, however, limited to single-celled organisms. Most plants have the ability to reproduce asexually. Sexual reproduction requires the involvement of two individuals, typically one of each sex.

Asexual Reproduction

Asexual reproduction is the process by which an organism creates a genetically-similar or identical copy of itself without a contribution of genetic material from another individual. Bacteria divide asexually via binary fission; viruses take control of host cells to produce more viruses; Hydras (invertebrates of the order *Hydroidea*) and yeasts are able to reproduce by budding. These organisms do not have different sexes, and they are capable of "splitting" themselves into two or more individuals. Some 'asexual' species, like hydra and jellyfish, may also reproduce sexually. For instance, most plants are capable of vegetative reproduction—reproduction without seeds or spores—but can also reproduce sexually. Likewise, bacteria may exchange genetic information by conjugation. Other ways of asexual reproduction include parthogenesis, fragmentation and spore formation that involves only mitosis. Parthenogenesis is the growth and development of embryo or seed without fertilization by a male. Parthenogenesis occurs naturally in some species, including lower plants (where it is called apomixis), invertebrates (e.g. water fleas, aphids, some bees and parasitic wasps), and vertebrates (e.g. some reptiles, fish, and, very rarely, birds and sharks). It is sometimes also used to describe reproduction modes in hermaphroditic species which can self-fertilize.

Sexual Reproduction

Sexual reproduction is a biological process by which organisms create descendants that have a combination of genetic material contributed from two (usually) different members of the species. Each of two parent organisms contributes half of the offspring's genetic makeup by creating haploid gametes. Most organisms form two different types of gametes. In these *anisogamous* species, the two sexes are referred to as male (producing sperm or microspores) and female (producing ova or megaspores). In *isogamous species* the gametes are similar or identical in form, but may have separable properties and then may be given other different names. For example, in the green alga, *Chlamydomonas reinhardtii*, there are so-called "plus" and "minus" gametes. A few types of organisms, such as ciliates, have more than two kinds of gametes.

Most animals (including humans) and plants reproduce sexually. Sexually reproducing organisms have two sets of genes for every trait (called alleles). Offspring inherit one allele for each trait from each parent, thereby ensuring that offspring have a combination of the parents' genes. Having two copies of every gene, only one of which is expressed, allows deleterious alleles to be masked, an advantage believed to have led to the evolutionary development of diploidy (Otto and Goldstein).

Autogamy

Self-fertilization (also known as autogamy) occurs in hermaphroditic organisms where the two gametes fused in fertilization come from the same individual. They are bound and all the cells merge to form one new gamete.

Mitosis and Meiosis

Mitosis and meiosis are an integral part of cell division. Mitosis occurs in somatic cells, while meiosis occurs in gametes. Mitosis The resultant number of cells in mitosis is twice the number of original cells. The number of chromosomes in the daughter cells is the same as that of the parent cell.

Meiosis The resultant number of cells is four times the number of original cells. This results in cells with half the number of chromosomes present in the parent cell. A diploid cell duplicates itself, then undergoes two divisions (tetraploid to diploid to haploid), in the process forming four haploid cells. This process occurs in two phases, meiosis I and meiosis II.

Same-sex Reproduction

In recent decades, developmental biologists have been researching and developing techniques to facilitate same-sex reproduction. The obvious approaches, subject to a growing amount of activity, are female sperm and male eggs, with female sperm closer to being a reality for humans, given that Japanese scientists have already created female sperm for chickens. More recently, by altering the function of a few genes involved with imprinting, other Japanese scientists combined two mouse eggs to produce daughter mice.

Reproductive Strategies

There are a wide range of reproductive strategies employed by different species. Some animals, such as the human and Northern Gannet, do not reach sexual maturity for many years after birth and even then produce few offspring. Others reproduce quickly; but, under normal circumstances, most offspring do not survive to adulthood. For example, a rabbit (mature after 8 months) can produce 10–30 offspring per year, and a fruit fly (mature after 10–14 days) can produce up to 900 offspring per year. These two main strategies are known as K-selection (few offspring) and r-selection (many offspring). Which strategy is favoured by evolution depends on a variety of circumstances. Animals with few offspring can devote more resources to the nurturing and protection of each individual offspring, thus reducing the need for many offspring. On the other hand, animals with many offspring may devote fewer resources to each individual offspring; for these types of animals it is common for many offspring to die soon after birth, but enough individuals typically survive to maintain the population.

Other Types of Reproductive Strategies

Polycyclic animals reproduce intermittently throughout their lives. Semelparous organisms reproduce only once in their lifetime, such as annual plants. Often, they die shortly after reproduction. This is a characteristic of r-strategists. Iteroparous organisms produce offspring in successive (e.g. annual or seasonal) cycles, such as perennial plants. Iteroparous animals survive over multiple seasons (or periodic condition changes). This is a characteristic of K-strategists.

Asexual vs. Sexual Reproduction

Organisms that reproduce through asexual reproduction tend to grow in number exponentially. However, because they rely on mutation for variations in their DNA, all members of the species have similar vulnerabilities. Organisms that reproduce sexually yield a smaller number of offspring, but the large amount of variation in their genes makes them less susceptible to disease. Many organisms can reproduce sexually as well as asexually. Aphids, slime molds, sea anemones, some species of starfish (by fragmentation), and many plants are examples. When environmental factors are favourable, asexual reproduction is employed to exploit sui conditions for survival such as an abundant food supply, adequate shelter, favorable climate, disease, optimum pH or a proper mix of other lifestyle requirements. Populations of these organisms increase exponentially via asexual reproductive strategies to take full advantage of the rich supply resources.

When food sources have been depleted, the climate becomes hostile, or individual survival is jeopardized by some other adverse change in living conditions, these organisms switch to sexual forms of reproduction. Sexual reproduction ensures a mixing of the gene pool of the species.

The variations found in offspring of sexual reproduction allow some individuals to be better suited for survival and provide a mechanism for selective adaptation to occur. In addition, sexual reproduction usually results in the formation of a life stage that is able to endure the conditions that threaten the offspring of an asexual parent. Thus, seeds, spores, eggs, pupae, cysts or other "over-wintering" stages of sexual reproduction ensure the survival during unfavourable times and the organism can "wait out" adverse situations until a swing back to suitability occurs.

Life Without Reproduction

The existence of life without reproduction is the subject of some speculation. The biological study of how the origin of life led from non-reproducing elements to reproducing organisms is called abiogenesis. Whether or not there were several independent abiogenetic events, biologists believe that the last universal ancestor to all present life on earth lived about 3.5 billion years ago.

Today, some scientists have speculated about the possibility of creating life non-reproductively in the laboratory. Several scientists have succeeded in producing simple viruses from entirely non-living materials. The virus is often regarded as not alive. Being nothing more than a bit of RNA or DNA in a protein capsule, they have no metabolism and can only replicate with the assistance of a hijacked cell's metabolic machinery. The production of a truly living organism (*e.g.*, a simple bacterium) with no ancestors would be a much more complex task, but may well be possible according to current biological knowledge.

Lottery Principle

Sexual reproduction has many drawbacks, since it requires far more energy than asexual reproduction and diverts the organisms from other pursuits, and there is some argument about why so many species use it. George C. Williams used lottery tickets as an analogy in one explanation for the widespread use of sexual reproduction. He argued that asexual reproduction, which produces little or no genetic variety in offspring, was like buying many tickets that all have the same number, limiting the chance of "winning"-that is, producing surviving offspring. Sexual reproduction, he argued, was like purchasing fewer tickets but with a greater variety of numbers and therefore a greater chance of success.

The point of this analogy is that since asexual reproduction does not produce genetic variations, there is little ability to quickly adapt to a changing environment. The lottery principle is less accepted these days because of evidence that asexual reproduction is more prevalent in unstable environments, the opposite of what it predicts.

Plant Reproduction

Plant reproduction is the production of new individuals or offspring in plants, which can be accomplished by both sexual or asexual means. Sexual reproduction produces offspring by the fusion of gametes, resulting in offspring genetically different from the parent or parents. Asexual reproduction produces new individuals without the fusion of gametes, they are genetically identical to the parent plants and each other, except when mutations occur. In seed plants, the offspring are packaged in a protective seed, which is used as an agent of dispersal.

Asexual Reproduction

Plants have two main types of asexual reproduction in which new plants are produced that are genetically identical clones of the parent individual. "Vegetative" reproduction involves a vegetative piece of the original plant (budding, tillering, etc.) and is distinguished from "apomixis", which is a "replacement" for sexual reproduction, and in some cases involves seeds. Apomixis occurs in many plant species and also in some non-plant organisms. For apomixis and similar processes in non-plant organisms, see parthenogenesis.

Natural vegetative reproduction is mostly a process found in herbaceous and woody perennial plants, and typically involves structural modifications of the stem or roots and in a few species leaves. Most plant species that employ vegetative reproduction, do so as a means to perennialize the plants, allowing them to survive from one season to the next and often facilitating their expansion in size. A plant that persists in a location through vegetative reproduction of individuals constitutes a clonal colony, a single ramet, or apparent individual, of a clonal colony is genetically identical to all others in the same colony. The distance that a plant can move during vegetative reproduction is limited, though some plants can produce ramets from branching rhizomes or stolons that cover a wide area, often in only a few growing seasons. In a sense, this process is not one of "reproduction" but one of survival and expansion of biomass of the individual. When an individual organism increases in size via cell multiplication and remains intact, the process is called "vegetative growth".

However, in vegetative reproduction, the new plants that result are new individuals in almost every respect except genetic. A major disadvantage to vegetative reproduction, is the transmission of pathogens from parent to daughter plants; it is uncommon for pathogens to be transmitted from the plant to its seeds, though there are occasions when it occurs. Seeds generated by apomixis are a means of asexual reproduction, involving the formation and dispersal of seeds that do not originate from the fertilization of the embryos. Hawkweed (*Hieracium*), dandelion (*Taraxacum*), some Citrus (*Citrus*) and Kentucky blue grass (*Poa pratensis*) all use this form of asexual reproduction. Pseudogamy occurs in some plants that have apomictic seeds, where pollination is often needed to initiate embryo growth, though the pollen contributes no genetic material to the developing offspring. Other forms of apomixis occur in plants also, including the generation of a plantlet in replacement of a seed or the generation of bulbils instead of flowers, where new cloned individuals are produced.

Natural Vegetative Structures

The rhizome is a modified underground stem serving as an organ of vegetative reproduction, e.g. Polypody, Iris, Couch Grass and Nettles. Prostrate aerial stems, called runners or stolons are important vegetative reproduction organs in some species, such as the strawberry, numerous grasses, and some ferns. Adventitious buds form on roots near the ground surface, on damaged stems (as on the stumps of cut trees), or on old roots. These develop into above-ground stems and leaves.

A form of budding called suckering is the reproduction or regeneration of a plant by shoots that arise from an existing root system. Species that characteristically produce suckers include Elm (*Ulmus*), Dandelion (*Taraxacum*), and members of the Rose Family (*Rosa*). Another type of a vegetative reproduction is the production of bulbs. Plants like onion (*Allium cepa*), hyacinth (*Hyacinth*), narcissus (*Narcissus*) and tulips (*Tulipa*) reproduce by forming bulbs. Other plants like potatoes (*Solanum tuberosum*) and dahlia (*Dahlia*) reproduce by a method similar to bulbs: they produce tubers. Gladioli and crocuses (*Crocus*) reproduce by forming a bulb-like structure called a corm.

Human uses of Asexual Reproduction

The most common form of plant reproduction utilized by people is seeds, but a number of asexual methods are utilized which are usually enhancements of natural processes, including: cutting, grafting, budding, layering, division, sectioning of rhizomes or roots, stolons, tillers (suckers) and artificial propagation by laboratory tissue cloning. Asexual methods are most often used to propagate cultivars with individual desirable characteristics that do not come true from seed. Fruit tree propagation is frequently performed by budding or grafting desirable cultivars (clones), onto rootstocks that are also clones, propagated by layering.

In horticulture, a "cutting" is a branch that has been cut off from a mother plant below an internode and then rooted, often with the help of a rooting liquid or powder containing hormones. When a full root has formed and leaves begin to sprout anew, the clone is a self-sufficient plant, genetically identical to the mother plant. Examples include cuttings from the stems of blackberries (*Rubus occidentalis*), African violets (*Saintpaulia*), verbenas (*Verbena*) to produce new plants. A related use of cuttings is grafting, where a stem or bud is joined onto a different stem. Nurseries offer for sale trees with grafted stems that can produce four or more varieties of related fruits, including apples. The most common usage of grafting is the propagation of cultivars onto already rooted plants, sometimes the rootstock is used to dwarf the plants or protect them from root damaging pathogens.

Since vegetatively propagated plants are clones, they are important tools in plant research. When a clone is grown in various conditions, differences in growth can be ascribes to environmental effects instead of genetic differences.

Sexual Reproduction

Sexual reproduction involves two fundamental processes, meiosis which rearranges the genes and reduces the number of chromosomes, and fusion of gametes which restores the chromosome to a complete diploid number. In between these two processes, different types of plants vary. In plants and algae that undergo alternation of generations, a gametophyte is the multicellular structure, or phase, that is haploid, containing a single set of chromosomes:

The gametophyte produces male or female gametes (or both), by a process of cell division called mitosis. The fusion of male and female gametes produces a diploid zygote, which develops by repeated mitotic cell divisions into a multicellular sporophyte. Because the sporophyte is the product of the fusion of two haploid gametes, its cells are diploid, containing two sets of chromosomes. The mature sporophyte produces spores by a process called meiosis, sometimes referred to as "reduction division" because the chromosome pairs are separated once again to form single sets. The spores are therefore once again haploid and develop into a haploid gametophyte. In land plants such as ferns, mosses and liverworts the gametophyte is very small, as in ferns and their relatives. In flowering plants (angiosperms) It is reduced to only a few cells, where the female gametophyte (embryo sac) is known as a megagametophyte and the male gametophyte (pollen) is called a micro gametophyte.

History of Sexual Reproduction

Unlike animals, plants are immobile, and cannot seek out sexual partners for reproduction. In the evolution of early plants, abiotic means, including water and wind, transported sperm for reproduction. The first plants were aquatic and released sperm freely into the water to be carried with the currents. Primitive land plants like liverworts and mosses had motile sperm that swam in a thin film of water or were splashed in water droplets from the male reproduction organs onto the female organs. As taller and more complex plants evolved, modifications in the alternation of generations evolved; in the Paleozoic era progymnosperms reproduced by using spores dispersed on the wind.

The seed plants including seed ferns, conifers and cordaites, which were all gymnosperms, evolved 350 million years ago; they had pollen grains that contained the male gametes for protection of the sperm during the process of transfer from the male to female parts. It is believed that insects fed on the pollen, and plants thus evolved to use insects to actively carry pollen from one plant to the next. Seed producing plants, which include the angiosperms and the gymnosperms, have heteromorphic alternation of generations with large sporophytes containing much reduced gametophytes. Angiosperms have distinctive reproductive organs called flowers, with carpels, and the female gametophyte is greatly reduced to a female embryo sac, with as few as eight cells. The male gametophyte consists of the pollen grains. The sperm of seed plants are non-motile, except for two older groups of plants, the Cycadophyta and the Ginkgophyta, which have flagellated sperm.

Flowering Plants

Flowering plants are the dominant plant form on land and they reproduce by sexual and asexual means. Often their most distinguishing feature is their reproductive organs, commonly called flowers. Sexual reproduction in flowering plants involves the production of male and female gametes, the transfer of the male gametes to the female ovules in a process called pollination. After pollination occurs, fertilization happens and the ovules grow into seeds with in a fruit. After the seeds are ready for dispersal, the fruit ripens and by various means the seeds are freed from the fruit and after varying amounts of time and under specific conditions the seeds germinate and grow into the next generation. The anther produces male gametophytes, the sperm is produced in pollen grains, which attach to the stigma on top of a carpel, in which the female gametophytes (inside ovules) are located. After the pollen tube grows through the carpel's style, the sex cell nuclei from the pollen grain migrate into the ovule to fertilize the egg cell and endosperm nuclei within the female gametophyte in a process termed double fertilization.

The resulting zygote develops into an embryo, while the triploid endosperm (one sperm cell plus two female cells) and female tissues of the ovule give rise to the surrounding tissues in the developing seed. The ovary, which produced the female gametophyte(s), then grows into a fruit, which surrounds the seed(s). Plants may either self-pollinate or cross-pollinate. Nonflowering plants like ferns, moss and liverworts use other means of sexual reproduction.

Adaptations

Flowers of wind pollinated plants tend to lack petals and or sepals. Typically large amounts of pollen are produced and pollination often occurs early in the growing season before leaves can interfere with the dispersal of the pollen. Many trees and all grasses and sedges are wind pollinated, as such they have no need for large fancy flowers. In plants that use insects or other animals to move pollen from one flower to the next, plants have developed greatly modified flower parts to attract pollinators and to facilitate the movement of pollen from one flower to the insect and from the insect back to the next flower.

Plants have a number of different means to attract pollinators including colour, scent, heat, nectar glands, eatable pollen and flower shape. Along with modifications involving the above structures two other conditions play a very important role in the sexual reproduction of flowering plants, the first is timing of flowering and the other is the size or number of flowers produced. Often plant species have a few large, very showy flower while others produce many small flowers, often flowers are collected together into large inflorescences to maximize their visual effect, becoming more noticeable to passing pollinators. Flowers are attraction strategies and sexual expressions are functional strategies used to produce the next generation of plants, with pollinators and plants having co-evolved, often to some extraordinary degrees, very often rendering mutual benefit.

The largest family of flowering plants is the orchids (Orchidaceae), estimated by some specialists to include up to 35,000 species, which often have highly specialized flowers used to attract insects and facilitate pollination. The stamens are modified to produce pollen in clusters called pollinium, which are attached to insects when crawling into the flower. The flower shapes are modified to force insects to pass by the pollen, which is "glued" to the insect. Some orchids are even more highly specialized, with flower shapes that mimic the shape of insects to attract them to 'mate' with the flowers, a few even have scents that mimic insect pheromones. Another large group of flowering plants is the Asteraceae or sunflower family with close to 22,000 species, which also have highly modified inflorescences that are flowers collected together in heads composed of a composite of individual flowers called florets. Heads with florets of one sex, when the flowers are pistillate or functionally staminate, or made up of all bisexual florets, are called homogamous and can include discoid and liguliflorous type heads. Some radiate heads may be homogamous too. Plants with heads that have florets of two or more sexual forms are called heterogamous and include radiate and disciform head forms, though some radiate heads may be heterogamous too.

Ferns

Ferns typically produce large diploid sporophytes with rhizomes, roots and leaves; and on fertile leaves called sporangium, spores are produced. The spores are released and germinate to produce short, thin gametophytes that are typically heart shaped, small and green in colour. The gametophytes or thallus, produce both motile sperm in the antheridia and egg cells in separate archegonia. After rains or when dew deposits a film of water, the motile sperm are splashed away from the antheridia, which are normally produce on the top side of the thallus, and swim in the film of water to the antheridia where they fertilize the egg.

To promote out crossing or cross fertilization the sperm are released before the eggs are receptive of the sperm, making it more likely that the sperm will fertilize the eggs of different thallus. A zygote is formed after fertilization, which grows into a new sporophytic plant. The condition of having separate sporephyte and gametophyte plants is call alternation of generations. Other plants with similar reproductive means include the *Psilotum*, *Lycopodium*, *Selaginella* and *Equisetum*.

Bryophytes

The bryophytes, which include liverworts, hornworts and mosses, reproduce both sexually and vegetatively. The gametophyte is the most commonly known phase of the plant. An early developmental stage in the gametophyte of mosses (immediately following germination of the meiospore) is called the protonema. All are small plants found growing in moist locations and like ferns, have motile sperm with flagella and need water to facilitate sexual reproduction.

These plants start as a haploid spore that grows into the dominate form, which is a multicellular haploid body with leaf-like structures that photosynthesize. Haploid gametes are produced in antherida and archegonia by mitosis. The sperm released from the antherida respond to chemicals released by ripe archegonia and swim to them in a film of water and fertilize the egg cells thus producing a zygote. The zygote divides by mitotic division and grows into a sporophyte that is diploid. The multicellular diploid sporophyte produces structures called spore capsules, which are connected by seta to the archegonia. The spore capsules produce spores by meiosis, when ripe the capsules burst open and the spores are released. Bryophytes show considerable variation in their breeding structures and the above is a basic outline. Also in some species each plant is one sex while other species produce both sexes on the same plant.

Sexual Expression

Many plants have evolved a complex sexuality, which is expressed in different combinations of their reproductive organs. Some species have separate male and female individuals, but the majority of plants have both male and female parts in the same flower. Some plants change their gender expression depending on a number of factors like age, time of day, or because of environmental conditions. Plant sexuality also varies within different populations of some species.

Reproductive System

The reproductive system or genital system is a system of organs within an organism which work together for the purpose of reproduction. Many non-living substances such as fluids, hormones, and pheromones are also important accessories to the reproductive system. Unlike most organ systems, the sexes of differentiated species often have significant differences. These differences allow for a combination of genetic material between two individuals, which allows for the possibility of greater genetic fitness of the offspring. The major organs of the human reproductive system include the external genitalia (penis and vulva) as well as a number of internal organs including the gamete producing gonads (testicles and ovaries). Diseases of the human reproductive system are very common and widespread, particularly communicable sexually transmitted diseases.

Most other vertebrate animals have generally similar reproductive systems consisting of gonads, ducts, and openings. However, there is a great diversity of physical adaptations as well as reproductive strategies in every group of vertebrates.

Human Reproductive System

Human reproduction takes place as internal fertilization by sexual intercourse. During this process, the erect penis of the male is inserted into the female's vagina until the male ejaculates semen, which contains sperm, into the female's vagina. The sperm then travels through the vagina and cervix into the uterus or fallopian tubes for fertilization of the ovum. Upon successful fertilization and implantation, gestation of the foetus then occurs within the female's uterus for approximately nine months, this process is known as pregnancy in humans. Gestation ends with birth, the process of birth is known as labour. Labour consists of the muscles of the uterus contracting, the cervix dilating, and the baby passing out the vagina. Human's babies and children are nearly helpless and require high levels of parental care for many years. One important type of parental care is the use of the mammary glands in the female breasts to nurse the baby.

Humans have a high level of sexual differentiation. In addition to differences in nearly every reproductive organ, numerous differences typically occur in secondary sexual characteristics.

Male Reproductive System

Partially shaved erect male genitalia.

1. Testicles,
2. Epididymis,
3. Corpus cavernosa,
4. Foreskin,
5. Frenulum,
6. Urethral opening,
7. Glans penis,
8. Corpus spongiosum,
9. Penis,
10. Scrotum.

The human male reproductive system is a series of organs located outside of the body and around the pelvic region of a male that contribute towards the reproductive process. The primary direct function of the male reproductive system is to provide the male gamete or spermatozoa for fertilization of the ovum.

The major reproductive organs of the male can be grouped into three categories. The first category is sperm production and storage. Production takes place in the testes which are housed in the temperature regulating scrotum, immature sperm then travel to the epididymis for development and storage. The second category are the ejaculatory fluid producing glands which include the seminal vesicles, prostate, and the vas deferens.

The final category are those used for copulation, and deposition of the spermatozoa (sperm) within the female, these include the penis, urethra, vas deferens, and Cowper's gland. Major secondary sexual characteristics includes: larger, more muscular stature, deepened voice, facial and body hair, broad shoulders, and development of an adam's apple. An important sexual hormone of males is androgen, and particularly testosterone.

Female Reproductive System

Photograph of the vulva.

1. Pubic hair (shaved),
2. litoral hood,
3. Clitoris,
4. Labia majora,

5. Labia minora (enclosing the Vaginal Opening),
6. Perineum.

The human female reproductive system is a series of organs primarily located inside of the body and around the pelvic region of a female that contribute towards the reproductive process. The human female reproductive system contains three main parts: the vagina, which acts as the receptacle for the male's sperm, the uterus, which holds the developing fetus, and the ovaries, which produce the female's ova. The breasts are also an important reproductive organ during the parenting stage of reproduction.

The vagina meets the outside at the vulva, which also includes the labia, clitoris and urethra; during intercourse this area is lubricated by mucus secreted by the Bartholin's glands. The vagina is attached to the uterus through the cervix, while the uterus is attached to the ovaries via the fallopian tubes. At certain intervals, typically approximately every 28 days, the ovaries release an ovum, which passes through the fallopian tube into the uterus. The lining of the uterus, called the endometrium, and unfertilized ova are shed each cycle through a process known as menstruation. Major secondary sexual characteristics include: a smaller stature, a high percentage of body fat, wider hips, development of mammary glands, and enlargement of breasts. Important sexual hormones of females include estrogen and progesterone.

Production of Gametes

The production of gametes takes place within the gonads through a process known as gametogenesis. Gametogenesis occurs when certain types of germ cells undergo meiosis to split the normal diploid number of chromosomes in humans (n=46) into haploids cells containing only 23 chromosomes.

In males this process is known as spermatogenesis and takes place only after puberty in the seminiferous tubules of the testes. The immature spermatozoon or sperm are then sent to the epididymis where they gain a tail and motility. Each of the original diploid germs cells or primary spermatocytes forms four functional gametes which is each capable of fertilization. In females gametogenesis is known as oogenesis which occurs in the ovarian follicles of the ovaries. This process does not produce mature ovum until puberty. In contrast with males, each of the original diploid germ cells or primary oocytes will form only one mature ovum, and three polar bodies which are not capable of fertilization. It has long been understood that in females, unlike males, all of the primary oocytes ever found in a female will be created prior to birth, and that the final stages of ova production will then not resume until puberty. However, recent scientific data has challenged that hypothesis. This new data indicates that in at least some species of mammal oocytes continue to be replenished in females well after birth.

Development of the Reproductive System

The development of the reproductive system and urinary systems are closely tied in the development of the human fetus. Despite the differences between the adult male and female reproductive system, there are a number of homologous structures shared between them due to their common origins within the fetus. Both organ systems are derived from the intermediate mesoderm. The three main fetal precursors of the reproductive organs are the Wolffian duct, Müllerian ducts, and the gonad. Endocrine hormones are a well known and critical controlling factor in the normal differentiation of the reproductive system. The Wolffian duct forms the epididymis, vas deferns, ductus deferens, ejaculatory duct, and seminal vesicle in the male reproductive system and essentially disappears in the female reproductive system.

For the Müllerian Duct this process is reversed as it essentially disappears in the male reproductive system and forms the fallopian tubes, uterus, and vagina in the female system. In both sexes the gonad goes on to form the testes and ovaries, because they are derived from the same undeveloped structure they are considered homologous organs. There are a number of other homologous structures shared between male and female reproductive systems. However, despite the similarity in function of the female fallopian tubes and the male epididymis and vas deferens, they are not homologous but rather analogous structures as they arise from different fetal structures.

Table: *Examples of homologous human reproductive organs*

Male organ	Female organ	Shared function
Cowper's gland	Bartholin's glands	Lubrication secretions
Penis	Clitoris	Erectile tissue and sensation
Testes	Ovary	Gamete production
Prostate gland	Skene's gland	Ejaculatory fluid and sensation

Diseases of the Human Reproductive System

Like all complex organ systems the human reproductive system is affected by many diseases. There are four main categories of reproductive diseases in humans. They are:

1) genetic or congenital abnormalities,
2) cancers,
3) infections which are often sexually transmitted diseases, and
4) functional problems cause by environmental factors, physical damage, psychological issues, autoimmune disorders, or other causes.

The best known type of functional problems include sexual dysfunction and infertility which are both broad terms relating to many disorders with many causes. Specific reproductive diseases are often symptoms of other diseases and disorders, or have multiple, or unknown causes making them difficult to classify. Examples of unclassifiable disorders include Peyronie's disease in males and endometriosis in females. Many congenial conditions cause reproductive abnormalities but are better known for their other symptoms, these include: Turner syndrome, Klinefelter syndrome, Cystic fibrosis, and Bloom syndrome. It is also known that disruption of the endocrine system by certain chemical adversely affects the development of the reproductive system and can cause vaginal cancer. Many other reproductive diseases have also been link to exposure to synthetic and environmental chemicals. Common chemicals with known links to reproductive disorders include: lead, dioxin, styrene, toluene, and pesticides.

Examples of Congenital Abnormalities

- Kallmann syndrome-Genetic disorder causing decreased functioning of the sex hormone-producing glands caused by a deficiency of a hormone.
- Cryptorchidism-Absence of one or both testes from the scrotum.
- Androgen insensitivity syndrome-A genetic disorder causing people who are genetically male (i.e. XY chromosome pair) to develop sexually as a female due to an inability to utilize androgen.
- Intersexuality-A person who has genitalia and/or other sexual traits which are not clearly male or female.

Examples of Cancers

- Prostate cancer-Cancer of the prostate gland.
- Breast cancer-Cancer of the mammary gland.
- Ovarian cancer-Cancer of the ovary.
- Penile cancer-Cancer of penis.
- Uterine cancer-Cancer of the uterus.

- Testicular cancer-Cancer of the testicles.
- Cervical Cancer-Cancer of the cervix.

Examples of Infections

- HIV-Infection by the retrovirus known as human immunodeficiency virus.
- Genital warts-Sexually transmitted infection caused by some sub-types of human papillomavirus (HPV).
- Herpes simplex-Sexually transmitted infection caused by a virus called herpes simplex virus (HSV) type 2
- Gonorrhea-Common sexually transmitted disease caused by the Gram-negative bacterium *Neisseria gonorrheae*
- Yeast infection-Infection of the vagina by any species of the fungus genus *Candida*.
- Pelvic inflammatory disease-Painful infection of the female uterus, fallopian tubes, and/or ovaries with associated scar formation and adhesions to nearby tissues and organs.
- Syphilis-Sexually transmitted infection caused by the bacterium *Treponema pallidum*.
- Pubic lice-Infection of the pubic hair by crab lice, *Phthirius pubis*.
- Trichomoniasis-Sexually transmitted infection by the single-celled protozoan parasite *Trichomonas vaginalis*.

Examples of Functional Problems

- Impotence-The inability of a male to produce or maintain an erection.
- Hypogonadism-A lack of function of the gonads, in regards to either hormones or gamete production.
- Ectopic pregnancy-When a fertilized ovum is implanted in any tissue other than the uterine wall.
- Hypoactive sexual desire disorder-A low level of sexual desire and interest.
- Female sexual arousal disorder-A condition of decreased, insufficient, or absent lubrication in females during sexual activity
- Premature ejaculation-A lack of voluntary control over ejaculation.

Other Vertebrates

Vertebrate animals all share key elements of their reproductive systems. They all have gamete producing organs or gonads. These gonads are then connected by oviducts to an opening to the outside of the body, typically the cloaca, but sometime to a unique pore such as a vagina or intromittent organ.

Mammals

Most mammal reproductive systems are similar, however, there are some notable differences between the "normal" mammal and humans. For instance, most mammalian males have a penis which is stored internally until erect, and most have a penis bone or baculum. Additionally, males of most species do not remain continually sexually fertile as humans do. Like humans, most groups of mammals have descended testicles found within a scrotum, however, others have descended testicles that rest on the ventral body wall, and a few groups of mammals, such as elephants, have undescended testicles found deep within their body cavities near their kidneys.

Marsupials are unique in that the female has two vaginae, both of which open externally through one orifice but lead to different compartments within the uterus; males usually have a two-pronged penis which corresponds to the females' two vaginae. Marsupials typically develop their offspring in an external pouch containing teats to which their newborn young (joeys) attach themselves for post uterine development. Also, marsupials have a unique prepenial scrotum.

The uterus and vagina are unique to mammals with no homologue in birds, reptiles, amphibians, or fish. In place of the uterus the other vertebrate groups have an unmodified oviduct leading directly to a cloaca, which is a shared exit-hole for gametes, urine, and feces. Monotremes (i.e. platypus and echidnas), a group of egg-laying mammals, also lack a uterus and vagina, and in that respect have a reproductive system resembling that of a reptile.

Birds

Male and female birds have a cloaca, an opening through which eggs, sperm, and wastes pass. Intercourse is performed by pressing the lips of the cloacae together, which is sometimes known as the "cloacal kiss", during which time the male transfers his sperm to the female. A few species of birds (e.g. most waterfowl) have a intromittent organ which is known as a phallus that is analogous to the mammals' penis. The female lays amniotic eggs in which the young gestate. Unlike most vertebrates female birds typically have only one functional ovary and oviduct. As a group, birds, like mammals, are noted for their high level of parental care.

Reptiles

Reptiles are almost all sexually dimorphic, and exhibit internal fertilization through the cloaca. Some reptiles lay eggs while others are viviparous (animals that deliver live young). Reproductive organs are found within the cloaca of reptiles. Most male reptiles have copulatory organs, which are usually retracted or inverted and stored inside the body. In turtles and crocodilians, the male has a single median penis-like organ, while male snakes and lizards each possess a pair of penis-like organs.

Amphibians

Most amphibians exhibit external fertilization of eggs, typically within the water, though some amphibians such as caecilians have internal fertilization. All have paired, internal gonads, connected by ducts to the cloaca.

Fish

Fish exhibit a wide range of different reproductive strategies. Most fish however are oviparous and exhibit external fertilization. In this process, females use their cloaca to release a large quantities their gametes, called spawn, into the water and one or more males release "milt", a white fluid containing many sperm over the unfertilized eggs. Other species of fish are oviparous and have internal fertilization aided by pelvic or anal fins that are modified into an intromittent organ analogous to the human penis. A small portion of fish species are either viviparous or ovoviviparous, and are collectively known as livebearers. Fish gonads are typically pairs of either ovaries or testes. Most fish are sexually dimorphic but some species are hermaphroditic or unisexual.

Invertebrates

Invertebrates have an extremely diverse array of reproductive systems, the only commonality may be that they all lay eggs. Also, aside from cephalopods, and arthropods, nearly all other invertebrates are hermaphroditic and exhibit external fertilization.

Cephalopods

All cephalopods are sexually dimorphic and reproduce by laying eggs. Most cephalopods have semi-internal fertilization in which the male places his gametes inside the female's mantle cavity or pallial cavity to fertilize the ova found in the female's single ovary. Likewise, male cephalopods have only a single teste. In the female of most cephalopods the nidamental glands aid in development of the egg.

The "penis" in most unshelled male cephalopods (Coleoidea) is a long and muscular end of the gonoduct used to transfer spermatophores to a modified arm called a hectocotylus. That in turn is used to transfer the spermatophores to the female. In species where the hectocotylus is missing, the "penis" is long and able to extend beyond the mantle cavity and transfers the spermatophores directly to the female.Many cephalopods shed their gonads during reproduction, and thus only reproduce once. Most cephalopods die after reproducing. Females nautilus however, have the ability to regenerate their gonads, making them the only cephalopods to spawn once per year. The females in many cephalopod species exhibit some level of parental protection for their eggs.

Prototheria

Prototheria is a taxonomic group, or taxon, to which the order Monotremata belongs. It is conventionally ranked as a subclass within the mammals. Most of the animals in this group are extinct. The egg-laying monotremes are known from fossils of the Cretaceous and Cenozoic periods; they are represented today by the platypus and several species of echidna. The names Prototheria, Metatheria and Eutheria (meaning "first beasts", "changed beasts", and "true beasts") refer to the three mammalian groupings which have living representatives. Each of the three may be defined as a putative clade comprising a living crown-group (respectively the Monotremata, Marsupialia and Placentalia) plus any fossil species which are more closely related to that crown-group than to any other animals.

The threefold division of living mammals into monotremes, marsupials and placentals was already well established when Thomas Huxley proposed the names Metatheria and Eutheria to incorporate the two latter groups in 1880. Initially treated as subclasses, Metatheria and Eutheria are by convention now grouped as infraclasses of the subclass Theria, and in more recent proposals have been demoted further (to cohorts or even magnorders), as cladistic reappraisals of the relationships between living and fossil mammals have suggested that the Theria itself should be reduced in rank. Prototheria, on the other hand, was generally recognised as a subclass until quite recently, on the basis of an hypothesis which defined the group by two supposed synapomorphies:

(1) formation of the side wall of the braincase from a bone called the anterior lamina, contrasting with the alisphenoid in therians; and

(2) a linear alignment of molar cusps, contrasting with a triangular arrangement in therians.

These characters appeared to unite monotremes with a range of Mesozoic fossil orders (Morganucodonta, Triconodonta, Docodonta and Multituberculata) in a broader clade for which the name Prototheria was retained, and of which monotremes were thought to be only the last surviving branch.The evidence which was held to support this grouping is now universally discounted. In the first place, examination of embryos has revealed that the development of the braincase wall is essentially identical in therians and in 'prototherians': the anterior lamina simply fuses with the alisphenoid in therians, and therefore the 'prototherian' condition of the braincase wall is primitive for all mammals while the therian condition can be derived from it. Additionally, the linear alignment of molar cusps is also primitive for all mammals. Therefore, neither of these states can supply a uniquely shared derived character which would support a 'prototherian' grouping of orders in contradistinction to Theria.

In a further reappraisal, the molars of embryonic and fossil monotremes (living monotreme adults are toothless) appear to demonstrate an ancestral pattern of cusps which is similar to the triangular arrangement observed in therians. Some peculiarities of this dentition support an alternative grouping of monotremes with certain recently-discovered fossil forms into a proposed new clade known as the Australosphenida, and also suggest that the triangular array of cusps may have evolved independently in australosphenidans and therians.

The Australosphenida hypothesis remains controversial, and some taxonomists prefer to maintain the name Prototheria as a fitting contrast to the other group of living mammals, the Theria. In theory, the Prototheria is taxonomically redundant, since Monotremata is currently the only order which can still be confidently included, but its retention might be justified if new fossil evidence, or a re-examination of known fossils, enables extinct relatives of the monotremes to be identified and placed within a wider grouping.

Theria

Theria is a subclass of mammals that give birth to live young without using a shelled egg, including both eutherians (placental mammals) and metatherians (marsupials and their ancestors).

Extent

Members of this subclass have external ears, most can suckle on a nipple, and have an ankle specialized for power and range of motion.

Therians are often classified by their specialized dentition. Almost all currently extant (not extinct) mammals are therians. The only known exceptions are the platypus and the echidnas (spiny anteater), both of which are prototherian monotremes.

Taxonomy

The rank of 'Theria' may vary depending on the classification system used. The textbook classification system by Vaughan et al. (2000) gives the following:

Class Mammalia

- Subclass Theria: live-bearing mammals;
  - Infraclass Metatheria: marsupials
  - Infraclass Eutheria: placentals.

In the above system Theria is a subclass. Alternatively, in the system proposed by McKenna and Bell (1997) it is ranked as a supercohort under the subclass Theriiformes:

Class Mammalia

- Subclass Theriiformes: live-bearing mammals and their prehistoric relatives;
  - Infraclass Holotheria: modern live-bearing mammals and their prehistoric relatives.
    - Legion Cladotheria
    - Sublegion Zatheria
    - Infralegion Tribosphenida
    - Supercohort Theria: therian mammals
    - Cohort Marsupialia: marsupials
    - Cohort Placentalia: placentals.

Another classification proposed by Luo et al. (2002) does not assign any rank to the taxonomic levels, but uses a purely cladistic system instead.

Marsupial

Marsupials are an infraclass of mammals, characterized by a distinctive pouch (called the *marsupium*), in which females carry their young through early infancy.

History

It was once commonly believed that marsupials were a primitive forerunner of modern placental mammals, but fossil evidence, first presented by researcher M.J. Spechtt in 1982, conflicts with this assumption. Instead, both main branches of the mammal tree appear to have evolved concurrently toward the end of the Mesozoic era.

In the absence of soft tissues, such as the pouch and reproductive system, fossil marsupials can be distinguished from placentals by the form of their teeth; primitive marsupials possess four pairs of molar teeth in each jaw, whereas placental mammals never have more than three pairs. Using this criterion, the earliest known marsupial is *Sinodelphys szalayi*, which lived in China around 125 million years ago. This makes it almost contemporary to the earliest placental fossils, which have been found in the same area.

The discovery of Chinese marsupials appears to support the idea that marsupials reached Australia via Southeast Asia. There are a few species of marsupials still living in Asia, especially in the Sulawesi region of Indonesia. These marsupials coexist with primates, hooved mammals and other placentals. However, due to the fact that Australia and China were separated by the wide Tethys Sea in the early Cretaceous into the Northern continent of Laurasia and Southern continent of Gondwana, marsupials had to take a much longer route around. From their origin in East Laurasia (modern day China), they spread westwards into modern North America (still attached to Eurasia) and skipped across to South America, which was connected to North America up until around 65MYA.

Here they radiated into Borhyaenids and Shrew Opossums, creating a unique fauna found in South America and Antarctica (which were connected until 35MYA). Marsupials reached Australia via Antarctica about 50MYA just after Australia had split off, suggesting a single dispersion event of several of just one species, related to South America's Monito del Monte (Microbiothere), rafted across the widening, but still narrow gap between Australia and Antarctica at that time. In Australia, being the only mammals present (except a few Austrosphenids like echidnas and platypuses) they radiated into the wide varieties we see today, even island hopping some way through the Indonesian archipelagos, almost completing a circumnavigation back to their homeland in China.

On most continents, placental mammals were much more successful and no marsupials survived, though in South America the opossums retained a strong presence, and the Tertiary saw the genesis of marsupial predators such as the borhyaenids and the saber-toothed *Thylacosmilus*. In Australia, however, marsupials displaced placental mammals entirely, and have since dominated the Australian ecosystem.

Marsupial success over placental mammals in Australia has been attributed to their comparatively low metabolic rate, a trait which would prove helpful in the hot Australian climate. As a result, native Australian placental mammals (such as hopping mice) are more recent immigrants.

Description

An early birth removes a developing marsupial from its parent's body much sooner than in placental mammals, and thus marsupials have not developed a complex placenta to protect the embryo from its mother's immune system. Though early birth places the tiny newborn marsupial at a greater environmental risk, it significantly reduces the dangers associated with long pregnancies, as there is no need to carry a large fetus to full-term in bad seasons. Because newborn marsupials must climb up to their mother's nipples, their front limbs are much more developed than the rest of the body at the time of birth. It is possible that this requirement has resulted in the limited range of locomotor adaptations in marsupials compared to placentals. Marsupials must develop a grasping forepaw during their early youth, making the transition from this limb into a hoof, wing, or flipper, as some groups of placental mammals have done, far more difficult.

There are about 334 species of marsupial, and over 200 are native to Australia and neighbouring northern islands. There are also 100 extant American species; these are centred mostly in South America, but the Great American Interchange has provided Central America with 13 species, and North America with one (the Virginia Opossum). Some structural features can be found in marsupials. Ossified patellae are absent. Marsupials (and also monotremes) also lack a gross communication (corpus callosum) between the right and left brain hemisphere.

Reproductive System

Marsupials' reproductive systems differ markedly from those of placental mammals (*Placentalia*). Females have two lateral vaginas, which lead to separate uteruses, but both open externally through the same orifice. A third canal, the median vagina, is used for birth. This canal can be transitory or permanent. The males generally have a two-pronged penis, which corresponds to the females' two vaginas. The penis is used only for discharging semen into females, and there is instead a urogenital sac used to store waste before expulsion.

Pregnant females develop a kind of yolk sac in their wombs, which delivers nutrients to the embryo. Marsupials give birth at a very early stage of development (about 4–5 weeks); after birth, newborn marsupials crawl up the bodies of their mothers and attach themselves to a nipple, which is located on the underside of the mother either inside a pouch called the *marsupium* or open to the environment. To crawl to the nipple and attach to it the marsupial must have well developed forelimbs and facial structures. This is accomplished by accelerating forelimb and facial development in marsupials compared to placental mammals. A result there is decelerated development of such structures as the hindlimb and brain. There they remain for a number of weeks, attached to the nipple. The offspring are eventually able to leave the *marsupium* for short periods, returning to it for warmth, protection and nourishment.

Taxonomy

Taxonomically, there are two primary divisions of Marsupialia: American marsupials and the Australian marsupials. The Order Microbiotheria (which has only one species, the Monito del Monte) is found in South America but is believed to be more closely related to the Australian marsupials. There are many small arboreal species in each group.

The term *opossums* is properly used to refer to the American species (though *possum* is a common diminutive), while similar Australian species are properly called *possums*.

The Virginia Opossum (*Didelphis virginiana*), the only North American marsupial north of Mexico.

- Superorder Ameridelphia;
  - o Order Didelphimorphia (93 species)
    - * Family Didelphidae: opossums.
  - o Order Paucituberculata (6 species)
    - * Family Caenolestidae: shrew opossums.
- Superorder Australidelphia
  - o Order †Yalkaparidontia
  - o Order Microbiotheria (1 species)
    - * Family Microbiotheriidae: Monito del Monte.
  - o Order Dasyuromorphia (71 species)
    - * Family †Thylacinidae: Thylacine
    - * Family Dasyuridae: antechinuses, quolls, dunnarts, Tasmanian Devil, and relatives
    - * Family Myrmecobiidae: Numbat.

- o Order Peramelemorphia (24 species)
  - * Family Thylacomyidae: bilbies
  - * Family †Chaeropodidae: Pig-footed Bandicoot
  - * Family Peramelidae: bandicoots and allies.
- o Order Notoryctemorphia (2 species)
  - * Family Notoryctidae: marsupial moles .
- o Order Diprotodontia (137 species)
  - * Family Phascolarctidae: Koala
  - * Family Vombatidae: wombats
  - * Family †Diprotodontidae: diprotodon
  - * Family Phalangeridae: brushtail possums and cuscuses
  - * Family Burramyidae: pygmy possums
  - * Family Tarsipedidae: Honey Possum
  - * Family Petauridae: Striped Possum, Leadbeater's Possum, Yellow-bellied Glider, Sugar Glider, Mahogany Glider, Squirrel Glider
  - * Family Pseudocheiridae: ringtailed possums and relatives
  - * Family Potoridae: potoroos, rat kangaroos, bettongs
  - * Family Acrobatidae: Feathertail Glider and Feather-tailed Possum
  - * Family Hypsiprymnodontidae: Musky Rat-kangaroo
  - * Family Macropodidae: kangaroos, wallabies, and relatives
  - * Family †Thylacoleonidae: marsupial lions.
- o Order †Sparassodonta.

Mammaliaformes

Mammaliaformes ("mammal-shaped") is a clade that contains the mammals and their closest extinct relatives. Phylogenetically it is defined as a clade including the most recent common ancestor of *Sinoconodon*, morganuconodonts, docodonts, Monotremata, Marsupialia, Placentalia, extinct members of this clade, and all of its descendants.

The precise phylogeny is disputed due to the scantness of evidence in the fossil record. However, it is thought that the Mammaliaformes were of three major groups: Allotheria, the longest extinct lineage of pre-mammals; Docodonta, including close relatives such as Morganucodonta; and Symmetrodonta, the most basal of modern mammals. Mammaliaformes radiated from Cynodontia. The Probainognathians of the Eucynodont clade probably evolved into the early mammaliaformes, but the branch Allotheria was so different that they may have come from an entirely different group of cynodonts.

Early mammaliforms were generally shrew-like in appearance and size, and most of their distinguishing characteristics were internal. In particular, the structure of the mammaliform (and mammal) jaw and arrangement of teeth is nearly unique. Instead of having many teeth that are frequently replaced, mammals have one set of baby teeth and later one set of adult teeth which fit together precisely. This is thought to aid in the grinding of food to make it quicker to digest. Being warm-blooded requires more calories than "cold-blooded" animals, so quickening the pace of digestion is a necessity. The drawback to the fixed dentition is that worn teeth cannot be replaced, as was possible for the reptilian ancestors of mammaliforms. However, small mammals generally being very short-lived compared to reptiles of the same size, this was not much of a problem during the early phase of their evolution, in which the trait was set. Early mammaliaformes were probably nocturnal. Mammaliforms have several common structures. Most importantly, mammaliforms have highly specialized molars, with cusps and flat regions for grinding food. This system is also unique to mammals, although it seems to have evolved convergently in pre-mammals multiple times.

Lactation and fur, along with other characteristically mammalian features, are also thought to characterize the Mammaliaformes, but these traits are difficult to study in the fossil record. The fossilized remains of *Castorocauda lutrasimilis* are a unique exception. Some non-mammal mammaliformes still retain reptile-like traits. Some mammaliformes had reptile-like locomotion. Furthermore, these mammaliformes still had some bones on their lower jaw seen in reptiles.

Monotreme

Monotremes are mammals that lay eggs (Prototheria) instead of giving birth to live young like marsupials (Metatheria) and placental mammals (Eutheria).

They are conventionally treated as comprising a single order Monotremata, though a recent classification proposes to divide them into the orders Platypoda (the Platypus along with its fossil relatives) and Tachyglossa (the echidnas, or spiny anteaters). The entire grouping is also traditionally placed into a subclass Prototheria, which was extended to include several fossil orders but these are no longer seen as constituting a natural group allied to monotreme ancestry. A controversial hypothesis now relates the monotremes to a different assemblage of fossil mammals in a clade termed Australosphenida.

General Characteristics

Like other mammals, monotremes are warm-blooded with a high metabolic rate (though not as high as other mammals, see below); have hair on their bodies; produce milk through mammary glands to feed their young; have a single bone in their lower jaw; and have three middle-ear bones. Monotremes were very poorly understood for many years, and to this day some of the 19th century myths that grew up around them endure. It is still sometimes thought, for example, that the monotremes are "inferior" or quasi-reptilian, and that they are a distant ancestor of the "superior" placental mammals. It now seems clear that modern monotremes are the survivors of an early branching of the mammal tree; a later branching is thought to have led to the marsupial and placental groups.

In common with reptiles and marsupials, monotremes lack the connective structure (corpus callosum) which allows communication between the right and left brain hemispheres in placentals.

The key anatomical difference between monotremes and other mammals is the one that gave them their name; *monotreme* means 'single opening' in Greek and comes from the fact that their urinary, defecatory, and reproductive systems all open into a single duct, the cloaca. This structure is very similar to the one found in reptiles. Monotremes and marsupials have a single cloaca (though marsupials also have a separate genital tract), while placental mammal females have separate openings for reproduction, urination, and defecation: the vagina, the urethra, and the anus.

Monotremes lay eggs. However, the egg is retained for some time within the mother, who actively provides the egg with nutrients. Monotremes also lactate, but have no defined nipples, excreting the milk from their mammary glands via openings in their skin. All species are long-lived, with low rates of reproduction and relatively prolonged parental care of infants. Infant echidnas are sometimes known as *puggles*, referencing their similarity in appearance to the Australian children's toy designed by Tony Barber. The same term, though not generally accepted, is popularly applied to young platypuses as well.

Extant monotremes lack teeth as adults. Fossil forms and modern platypus young have "tribosphenic" molars (with the occlusal surface formed by three cusps arranged in a triangle), which are one of the hallmarks of extant mammals. Some recent work suggests that monotremes acquired this form of molar *independently* of placental mammals and marsupials, although this is not well established. The jaw of monotremes is constructed somewhat differently from that of other mammals, and the jaw opening muscle is different. As in all true mammals, the tiny bones that conduct sound to the inner ear are fully incorporated into the skull, rather than lying in the jaw as in cynodonts and other pre-mammalian synapsids; this feature, too, is now claimed to have evolved independently in monotremes and therians, although, as with the analogous evolution of the tribosphenic molar, this is disputed. The external opening of the ear still lies at the base of the jaw. The imminent sequencing of the Platypus genome should shed light on this and many other questions regarding the evolutionary history of the monotremes.

The monotremes also have extra bones in the shoulder girdle, including an interclavicle and coracoid, which are not found in other mammals. Monotremes retain a reptile-like gait, with legs that are on the sides of rather than underneath the body. The monotreme leg bears a spur in the ankle region; the spur is non-functional in echidnas, but contains a powerful venom in the male Platypus.

Physiology

It is still sometimes said that monotremes have less developed internal temperature control mechanisms than other mammals, but recent research shows that monotremes maintain a constant body temperature in a wide variety of circumstances without difficulty (for example, the Platypus while living in an icy mountain stream). Early researchers were misled by two factors: firstly, monotremes maintain a lower average temperature than most mammals (around 32 °C (90 °F), compared to about 35 °C (95 °F) for marsupials, and 37 °C (99 °F) for most placentals); secondly, the Short-beaked Echidna (which is much easier to study than the reclusive Platypus) maintains normal temperature only when it is active: during cold weather, it conserves energy by "switching off" its temperature regulation. Finally, poor thermal regulation has also been observed in the hyraxes, which are placental mammals.

Their metabolic rate is remarkably low by mammalian standards. The Platypus has an average body temperature of about 32 °C (90 °F) rather than the 37 °C (99 °F) typical of placental mammals. Research suggests this has been a gradual adaptation to harsh environmental conditions on the part of the small number of surviving monotreme species rather than a historical characteristic of monotremes.

Contrary to previous research, the Echidna does enter REM sleep, albeit only when the ambient temperature of its environment is around 25 °C (77 °F). At the temperatures of 15 °C (59 °F) and 28 °C (82 °F), REMS is suppressed.

Taxonomy

The only surviving examples of monotremes are all indigenous to Australia and New Guinea, although there is evidence that they were once more widespread. Fossil and genetic evidence shows that the monotreme line diverged from other mammalian lines about 150 million years ago and that both the short-beaked and long-beaked echidna species are derived from a platypus-like ancestor. Fossils of a jaw fragment 110 million years old were found at Lightning Ridge, New South Wales.

These fragments, from species *Steropodon galmani*, are the oldest known fossils of monotremes. Fossils from the genera *Kollikodon*, *Teinolophos*, and *Obdurodon* have also been discovered. In 1991, a fossil tooth of a 61-million-year-old platypus was found in southern Argentina (since named *Monotrematum*, though it is now considered to be an *Obdurodon* species). Molecular clock and fossil dating suggest echidnas split from platypuses 19–48 million years ago.

- Order Monotremata;
  - o Suborder Platypoda.
    - * Family Ornithorhynchidae: platypus
    - * Genus *Ornithorhynchus*
    - * Platypus, *Ornithorhynchus anatinus*.
  - o Suborder Tachyglossa.
    - * Family Tachyglossidae: echidnas
    - * Genus *Tachyglossus*
    - * Short-beaked Echidna, *Tachyglossus aculeatus*
    - * *Tachyglossus aculeatus aculeatus*
    - * *Tachyglossus aculeatus acanthion*
    - * *Tachyglossus aculeatus lawesii*
    - * *Tachyglossus aculeatus multiaculeatus*
    - * *Tachyglossus aculeatus setosus*

* Genus *Zaglossus*
* Sir David's Long-beaked Echidna, *Zaglossus attenboroughi*
* Eastern Long-beaked Echidna, *Zaglossus bartoni*
* *Zaglossus bartoni bartoni*
* *Zaglossus bartoni clunius*
* *Zaglossus bartoni diamondi*
* *Zaglossus bartoni smeenki*
* Western Long-beaked Echidna, *Zaglossus bruijni.*

Fossil Monotremes

The fossil record of monotremes is relatively sparse. Although biochemical and anatomical evidence suggests that monotremes diverged from the mammalian lineage before the marsupials and placental mammals arose, only a handful of monotreme fossils are known from before the Miocene epoch. The few Mesozoic fossils that do exist, such as that of *Steropodon*, seem to indicate that the monotremes first evolved in Australia, during the Late Jurassic or Early Cretaceous. They subsequently spread to both South America and Antarctica, which were still united with Australia at that time, but may not have survived on either continent for long.

Chapter 2

Animal Breeding

Breeding is the producing of offspring, usually animals or plants. In animal breeding techniques such as inbreeding, line breeding and out crossing are utilized. In plant breeding similar methods are used. Charles Darwin discussed how selective breeding had been successful in producing change over time in his book, Origin of Species. Selective breeding was used by Darwin as a springboard to introduce the theory of natural selection, and to support it.

Inbreeding

Inbreeding is the reproduction from the mating of two genetically related parents, which can increase the chances of offspring being affected by recessive or deleterious traits. This generally leads to a decreased fitness of a population, which is called inbreeding depression. Deleterious alleles causing inbreeding depression can subsequently be removed through culling, which is also known as genetic purging. Livestock breeders often practice controlled breeding to eliminate undesirable characteristics within a population, which is also coupled with culling of what is considered unfit offspring, especially when trying to establish a new and desirable trait in the stock. In plant breeding, inbred lines are used as stocks for the creation of hybrid lines to make use of the heterosis effect. Inbreeding in plants also occurs naturally in the form of self-pollination.

Results

Inbreeding may result in a far higher phenotypic expression of deleterious recessive genes within a population than would normally be expected. As a result, first-generation inbred individuals are more likely to show physical and health defects, including:

- Reduced fertility both in litter size and sperm viability
- Increased genetic disorders
- Fluctuating facial asymmetry
- Lower birth rate
- Higher infant mortality
- Slower growth rate
- Smaller adult size
- Loss of immune system function.

Natural selection works to remove individuals who acquire the above types of traits from the gene pool. Therefore, many more individuals in the first generation of inbreeding will never live to reproduce. Over time, with isolation such as a population bottleneck caused by purposeful (assortative) breeding or natural environmental stresses, the deleterious inherited traits are culled. The cheetah once was reduced by disease, habitat restriction, overhunting of prey, competition from other predators (primarily lions, competition from human land use, etc.) to a very small number of individuals. All cheetahs now come from this very small gene pool. Should a virus appear that none of the cheetahs have resistance to, extinction is always a possibility. Currently, the threatening virus is feline infectious peritonitis, which has a disease rate in domestic cats from 1%-5%; in the cheetah population it is ranging between 50% to 60%. The cheetah is also known, in spite of its small gene pool, for few genetic illnesses.

Island species are often very inbred, as their isolation from the larger group on a mainland allows for natural selection to work upon their population. This type of isolation may result in the formation of race or even speciation, as the inbreeding first removes many deleterious genes, and allows expression of genes that allow a population to adapt to an ecosystem. As the adaptation becomes more pronounced the new species or race radiates from its entrance into the new space, or dies out if it cannot adapt and, most importantly, reproduce. The reduced genetic diversity that results from inbreeding may mean a species may not be able to adapt to changes in environmental conditions. Each individual will have similar immune systems, as immune systems are genetically based. Where a species becomes endangered, the population may fall below a minimum whereby the forced interbreeding between the remaining animals will result in extinction. In the South American sea lion, there was concern that recent population crashes would reduce genetic diversity. Historical analysis indicated that a population expansion from just two matrilineal lines were responsible for most individuals within the population. Even so, the diversity within the lines allowed for great variation in the gene pool that may inoculate the South American sea lion from extinction.

Natural breedings include inbreeding by necessity, and most animals only migrate when necessary. In many cases, the closest living mate is a mother, sister, grandmother, father, grandfather... In all cases the environment presents stresses to select or remove those individuals who cannot survive because of illness from the population.

In lions, prides are often followed by related males in bachelor groups. When the dominant male is killed or driven off by one of these bachelors, a father may be replaced with his son. There is no mechanism for preventing inbreeding or to ensure out crossing. In the prides, most lionesses are related to one another. If there is more than one dominant male, the group of alpha males are usually related. Two lines then are being "line bred". Also, in some populations such as the Crater lions, it is known that a population bottleneck has occurred. Researchers found far greater genetic heterozygosity than expected. In fact, predators are known for low genetic variance, along with most of the top portion of the tropic levels of an ecosystem. Additionally, the alpha males of two neighbouring prides can potentially be from the same litter; one brother may come to acquire leadership over another's pride, and subsequently mate with his 'nieces' or cousins. However, killing another male's cubs, upon the takeover, allows for the new selected gene complement of the incoming alpha male to prevail over the previous male. There are genetic assays being scheduled for lions to determine their genetic diversity. The preliminary studies show results inconsistent with the out crossing paradigm based on individual environments of the studied groups.

There was an assumption that wild populations do not inbreed; this is not what is observed in some cases in the wild. However, in species such as horses, animals in wild or feral conditions often drive off the young of both genders, thought to be a mechanism by which the species instinctively avoids some of the genetic consequences of inbreeding.

Calculation

The inbreeding is computed as a percentage of chances for two alleles to be identical by descent. This percentage is called "inbreeding coefficient". There are several methods to compute this percentage, the two main ways are the path method [1] and the tabular method.

Typical inbreeding percentages are as follows:

- Father/daughter – mother/son – brother/sister → 25%
- Half-brother/half-sister → 12.5%
- Uncle/niece – aunt/nephew → 12.5%
- Double first cousins → 12.5%
- Half-uncle/niece → 6.25%
- First cousins → 6.25%
- First cousins once removed – half-first cousins → 3.125%
- Second cousins – first cousins twice removed → 1.5625%
- Second cousins once removed – half-second cousins → .78125%.

An inbreeding calculation may be used to determine the general genetic distance among relatives by multiplying by 2, because any progeny would have a 1 in 2 risk of actually inheriting the identical alleles from both parents. For instance, the parent/child or sibling/sibling have 50% identical genetics.

NOTE: For siblings, the degree of genetic relationship is not an automatic 50% (as it is with parents and their children), but a range from 100% at one extreme – as in the case of identical twins (who obviously could not mate as they are the same sex) – to an exceedingly unlikely 0%. Siblings share an average of 50% of their genes, but unlike the 50% ratio between parents and children, the actual ratio between siblings in any given case can vary.

Domestic Animals

An intensive form of line breeding where an individual with highly desirable traits (S) is mated to his daughter (D1) and daughter's daughter (D2) and so on, in order to maximise the percentage of S's genes in the offsprings. The D3 offspring would have 87.5% of his genes while D4 offspring would have 93.75%. Such breeding methods can be used to create a "near clone" of a desirable individual. Breeding in domestic animals is assortative breeding primarily. Without the sorting of individuals by trait, a breed could not be established, nor could poor genetic material be removed.

Inbreeding is used by breeders of domestic animals to fix desirable genetic traits within a population or to attempt to remove deleterious traits by allowing them to manifest phenotypically from the genotypes. Inbreeding is defined as the use of close relations for breeding such as mother to son, father to daughter, brother to sister. Breeders must cull unfit breeding suppressed individuals and/or individuals who demonstrate either homozygosity or heterozygosity for genetic based diseases. The issue of casual breeders who inbreed irresponsibly is discussed in the following quotation on cattle:

Meanwhile, milk production per cow per lactation increased from 17,444 lbs to 25,013 lbs from 1978 to 1998 for the Holstein breed. Mean breeding values for milk of Holstein cows increased by 4,829 lbs during this period. High producing cows are increasingly difficult to breed and are subject to higher health costs than cows of lower genetic merit for production *(Cassell, 2001).* Intensive selection for higher yield has increased relationships among animals within breed and increased the rate of casual inbreeding. Many of the traits that affect profitability in crosses of modern dairy breeds have not been studied in designed experiments. Indeed, all crossbreeding research involving North American breeds and strains is very dated (McAllister, 2001) if it exists at all.

Line breeding, which is a milder form of inbreeding is accomplished through breeding of cousins, aunt to nephew, half brother to half sister. This was used to isolate breeds within the companion and livestock industry. For instance an animal with a desirable colour is bred back within the lines with identified selection traits whether it be milk production or adherence to breed standard of appearance or behaviour. Breeders must then cull unfit individuals, and in some cases the breeders will then outbreed to increase the level of genetic diversity. Again casual breeding is problematic as it is without the requisite culling of individuals who are either maladaptive, not to breed standard or carriers of poor genetic material that must be removed from a healthy breeding program.

Out crossing is where two unrelated individuals have been crossed to produce progeny. In out crossing, unless there is verifiable genetic information, one may find that all individuals are distantly related to an ancient progenitor. If the trait carries throughout a population, all individuals can have this trait.

This is called the founder's effect. In the well established breeds, that are commonly bred,a large gene pool is present. For example, in 2004, over 18,000 Persian cats were registered. A possibility exists for a complete outcross, if no barriers exist between the individuals to breed. However it is not always the case, and a form of distant line breeding occurs. Again it is up to the assortative breeder to know what sort of traits both positive and negative exist within the diversity of one breeding. This diversity of genetic expression, within even close relatives, increases the variability and diversity of viable stock.

The two dog sites above also point out that in the registered dog population, the onset of large numbers of casual breeders has corresponded with an increase in the number of genetic illnesses of dogs by not understanding how, why and which traits are inherited. The dog sites indicate that the largest percentage of dog breeders in the US are casual breeders. Therefore the investment in a papered animal, with an expected short term profit, motivates some to ignore the practice of culling. Casual breeders in companion animals often ignore breeding restrictions within their contracts with source companion animal breeders. The casual breeders breed the very culls that a genetics based breeder has released as a pet. The casual breeder also was cited in the quotes above on cattle raising.

Laboratory Animals

Systematic inbreeding and maintenance of inbred strains of laboratory mice and rats is of great importance for biomedical research. The inbreeding guarantees a consistent and uniform animal model for experimental purposes and enables genetic studies in congenic and knock-out animals. The use of inbred strains is also important for genetic studies in animal models, for example to distinguish genetic from environmental effects.

Evolution

Inbreeding has a variety of consequences. Allele exposure can cause genes to be expressed that are not otherwise expressed. This fact, combined with the fact that most mutations are recessive may indicate that inbreeding drives evolution. Speciation, a key process in evolution, depends on reproductive barriers, a necessary feature of which is inbreeding.

Line breeding

An intensive form of line breeding where an individual with highly desirable traits (S) is mated to his daughter (D1) and daughter's daughter (D2) and so on, in order to maximise the percentage of S's genes in the offsprings. The D3 offspring would have 87.5% of his genes while D4 offspring would have 93.75%. Such breeding methods can be used to create a "near clone" of a desirable individual. Line breeding is a form of inbreeding practiced by some animal breeders to "fix" desirable traits in a breed of animal, without as high a risk of producing undesirable traits that may occur with close inbreeding. A typical example of line breeding would be what in human parlance would be considered a mating of first cousins or more distantly related individuals who share a common ancestor. While line breeding is less likely to cause problems in the first generation than does inbreeding, over time, line breeding can reduce the genetic diversity of a population and cause problems related to a too-small genepool that may include an increased prevalence of genetic disorders and inbreeding depression.

Out crossing

Out crossing is the practice of introducing unrelated genetic material into a breeding line. It increases genetic diversity, thus reducing the probability of all individuals being subject to disease or reducing genetic abnormalities (only within the first generation). It actually can serve to increase the number of individuals who carry a disease recessively. It is used in line-breeding to restore vigour or size and fertility to a breeding line. "Line-breeding", is where animals carry a common ancestor in their pedigrees and are bred together, should be considered distinct from the term "in-breeding" which is the production of offspring by parents more closely related than the average.

Out crossing is now the norm of most purposeful breeding, contrary to what is commonly believed. The out crossing breeder intends to remove the traits by using "new blood". With dominant traits, one can still see the expression of the traits and can remove those traits whether one outcrosses, line breeds or inbreds. With recessives, out crossing allows for the recessive traits to migrate across a population. It may actually increase the number of individuals carrying a disease. The out crossing breeder then may have individuals that have many deleterious genes that are expressed by placing their animals against a similarly outcrossed individual. There is now a gamut of deleterious genes within each individual in many breeds. However one may increase the variance of genes within the gene pool by out crossing, protecting against extinction by a single stressor from the environment. In cats, there is currently a study running to determine the genetic diversity within the cat breeds.

Out crossing is believed to be the "norm" in the wild. However, it is not logical as migration occurs by necessity. Feral cats, for example are one of the most inbred as individuals remain nearby their original homes, unless environmental stresses drive them to migration. Breeders inbreed within their genetic pool, attempting to maintain desirable traits and to cull those traits that are undesirable. When undesirable traits begin to appear, breedings are selected to determine if a trait is recessive or dominant. Removal is accomplished by breeding two individuals of known genetic status, usually they are related. In nature, where breeding is not managed, out crossing rates may be estimated by genetic analysis, by employing mathematical models of mating systems such as the mixed mating model or the effective selfing model. This allows calculation of the amount of genetic exchange between populations, and thus provides insights into the biogeography and phytogeography of species.

Gregor Mendel used out crossing in his experiments with flowers for his breeding stock. He then used the resulting offspring to chart inheritance patterns, using the crossing of siblings, and backcrossing to parents to determine how inheritance functioned.

Farm Animal Breeding and Reproduction in Europe – Characteristics:

- Farm animal breeding includes all animal species bred for a wide range of purposes.
- For breeding, animals meeting defined criteria are selected from animal populations.
- Farm animal species have been selected for desirable traits since they were first domesticated.
- Efficient reproductive techniques, such as artificial insemination, allow genetic improvement to be rapidly disseminated from the top of the breeding pyramid to benefit all producers and society as a whole.
- Animal breeding and reproduction are most effective when incorporated into herd and population management strategies.
- Balanced breeding requires a long-term sustainable vision developed jointly by breeders, scientists, and society.
- The added value of investments in genetics is cumulative.
- Breeding is knowledge intensive.
- An international farmers' organisation has established guidelines for data collection, farm management, and genetic evaluation to assist farmers and farmers' organisations.
- Breeding is society sensitive because it drives changes in the genetic makeup of animals and the use of new technologies (e.g. genomics, computing sciences).
- Genomics opens innovative prospects for sustainable animal production.
- A Code of Good Practice for Farm Animal Breeding and Reproduction Organisations is in place to encourage transparency and a dialogue of breeders with society.

- Breeding supports the health, feed efficiency, and welfare of farm animals as well as adequate management of animals and the environment.

The domestication of livestock species some ten thousand years ago was a vital step in the development of human civilisation. Over the centuries, domestication evolved into breeding and the genetic improvement of livestock. Nowadays, breeders measure many different animal traits and choose the best animals to be the parents of the next generation. This leads to improvement generation after generation through the increased frequency of desired gene variants in the population. Because the breeding response is cumulative, permanent, and can be spread throughout the production chain, farm animal selection has a great impact on farm animal production.

Europe has always played an important role in improving the world's major livestock and aquaculture species. European breeds are used across the world, and European farmers and breeding organisations are major players on the global market. The European farm animal breeding sector will thus have a great influence on, and therefore responsibility for, the future genetic makeup and characteristics of farm animal populations worldwide.

A conservative estimate of the economic gain achieved each year by animal breeding at farm level is €1.83 billion in Europe alone. Hence, the genetic gain achieved by breeders is carried over to producers as an economic gain reaching approximately 1.5% of the economic value of EU farm animal production. Average annual economic gain from farm animal breeding (derived from improved production) in the EU/ in Europe

Europe	***(Mio €)***
Dairy cattle	430
Beef cattle	70
Pigs (Europe)	520
Broilers (Europe)a	610
Layers (Europe)a	125
Salmon, rainbow trout, seabass/Seabream, turbot	80
Total	1830

The Growing Demand for Food from Animals

There are some 6 billion people in the world today. Despite declining population growth rates, the world population is increasing by about 80 million a year –equivalent to the population of Germany. Around 95% of this increase is taking place in the developing world. The UN predicts that the world population will reach 9 billion by 2050. Also growing is the per capita demand for animal food, which should continue to rise for at least 20 years.

This so-called livestock revolution is a demand-driven evolution. The supply of cereals for human consumption should soon be sufficient to satisfy the demand in developing countries, but the supply of animal-derived foods is far from the mark. The 23% of people living in developed countries presently consume 3-4 times more meat and fish and 5-6 times more milk per capita than people in developing countries. As these poorer people get richer, one of the first things they want to buy is more nutritious and satisfying food, and this generally means more animal protein. Animal product consumption is thus increasing massively in developing countries, and will continue to do so over the next 15-20 years.

Developing countries are also playing a growing role in animal production. Southeast Asia, in recent decades, has tremendously stepped up its pork, poultry, egg, and aquaculture production. Other noteworthy examples are Brazil, for beef cattle, chicken, and animal feed, and Mexico and Argentina for beef cattle and Chile for salmon. Breeding activities are also on the rise in these countries. Both Asia and Europe are densely populated and have comparatively little land available for agriculture. In contrast, the Americas (and Oceania) have relatively large amounts of arable land and pasture as compared to population density. This creates for Europe and Asia a permanent risk of food dependence on the Americas. It also creates a 'natural' push to develop foods and food products for –and sell them to-the more densely populated areas. Related policies include farming subsidies in the USA, stimulation of genomics and biotechnology research, and a favourable business and social climate for developing and implementing new technologies.

Importance of Animal Production in the EU

Mankind has been adapting animals for the production of both food (meat, milk, eggs...) and non-food products (wool, leather, bones...) since domestication started. For some farmed fish and shellfish, domestication is still underway. Animal husbandry for other purposes is expected to become more important. Animals have cultural value; they are used for sports and as companions, for maintaining rural areas, and for medical applications (as models of human disease, potentially as providers of organs for transplantation). The value of animal production at farm level in the European Union-25 (EU25) is €132 billion, amounting to 40% of the value of agricultural production (2004). The EU25 counts some 6 million cattle (of which 23 million dairy cows), 103 million sheep, 12 million goats, 4.4 million horses, 11 million beehives, 151 million pigs, 670 million laying hens, 7368 million broilers, and 285 million turkeys (2003). In addition, the annual value of aquaculture business is €2.8 billion (EU25, 2003). The total arable production includes 400 million tonnes of animal feed production, and about 1/3 of the agricultural area is permanent grassland (EU15). Farm animals in the EU consume about 450 million tonnes of feed a year, of which 140 million tonnes are produced by the compound feed manufacturers. The turnover of the European feed industry is estimated at €35 billion. With EU enlargement, the number of farms has more than doubled (to 17 million), and the proportion of farmers in the workforce has grown from around 4% (EU15) to nearly 8% (EU25). The EU imports €66.6 billion worth of agricultural products and exports €55.7 billion, so the balance is negative. The role of animals for leisure and sports (e.g. dogs, horses) is increasing. In the EU15, over 1 million people get their income from horses.

Socio-Economic Aspects of Animal Breeding

Output at Farm Level

The €1.83-billion annual gain derived from breeding in Europe does not include the export of breeding stock (important in poultry, pigs, and salmon). If this is added, the estimated gain is easily twice as high. The costs associated with this economic gain are relatively small in relation to the improvement achieved, as genetic improvement is cumulative and investments in small populations at the top of the breeding pyramid are multiplied down the pyramid to the base level: animals for commercial production.

A Competitive Global Market

Farm animal breeding and reproduction operate in a competitive global market with very low margins. Changes in market shares can occur quickly, especially in areas where a handful of companies serve 90% of the global market. Small differences in performance, cost price, or knowledge implementation (e.g. exploiting genomics) may bring about such changes. As a result, smaller companies might not survive the competitive pressure in a fast-changing marketplace with evolving technologies. This applies notably to ruminant breeding, which is mainly still organized nationally. Here, cooperation between European players is required. Maintaining the strong position of breeding in

Europe is thus not automatic. It will require considerable efforts from breeders, but also the positive engagement of researchers, governments, funding bodies, legislative bodies, and European society as a whole.

A Knowledge-Intensive Sector

The farm-animal breeding and reproduction sector is knowledge intensive. This means not only that the sector exploits knowledge to provide the world with breeding stock, but also that the knowledge developed is or can be used to meet local, niche, or cultural demands inside or outside Europe. The past success of European breeding owes much to its longstanding close ties with universities and research institutes, fostering the dissemination of knowledge to the farm and individual breeder level. In some cases (e.g., dairy cattle) and some countries this dissemination has been very successful, but it still needs to be improved in order to take other situations and rapid changes in animal breeding and reproduction technology into account. The knowledge exploited includes not only genetics, genomics, physiology, reproduction, statistics, and computing science, but also ? importantly ??economics, animal husbandry, storage and analysis of large datasets, and efficient infrastructure design. Benefits of co-operation over species have been large. The aim is to achieve higher production levels and better product quality while improving animal health, reproduction, other physiological characteristics, and/or feed efficiency.

The reproductive characteristics of the animals used largely determine the spread of genetic progress and the organisation of breeding schemes. Breeding organisations in Europe spend around €150 million yearly on research, development, and implementation (in-house or outsourced to institutions such as universities). This includes money spent on estimating breeding values and executing breeding programmes, but not the cost of steps further in the chain, like multiplication and commercialisation. Actors outside Europe (in the USA, China, and Australasia, for instance) are making huge investments in genomics and other new biotechnologies. Research and business conditions are more favourable in the countries concerned than in Europe, and there tends to be a more positive attitude towards progress made possible by new technologies. Europe must also develop knowledge and understanding of new technologies with no immediate application, so that we are well placed to understand their benefits and risks. In developing these technologies, it is essential to ensure close collaboration between research and government as well as transparent inclusion of stakeholders from industry and society.

Emerging Technologies and Societal Concerns

Farm animal breeding operates in a sensitive area because it touches upon important issues: food, health, animals, selection, genes... This sensitivity is reflected in societal concerns related to food safety, ethics, cultural value, genetic diversity, animal welfare, and 'naturalness'. Animal breeders are actively seeking to address these concerns, both internally amongst specialists and through dialogue with policymakers and society. Here are some of the issues that they are addressing:

Sustainability: Breeders have a role to play in promoting sustainable worldwide animal production (a widely discussed concept which emerged in the last decade and was developed notably in the EU-funded SEFABAR project). To play this role optimally over the next two decades, the breeding sector must be both economically sound and attuned to societal needs and demands. This notably means taking into account aspects such as biodiversity, environmental protection, food quality and safety, animal health and welfare. European breeders have been addressing the ethics of animal breeding, notably by producing educational material on this topic and by developing ethical codes such as a Code of Good Practice for Farm Animal Breeding Organisations (Code-EFABAR). Centring on sustainability, this Code aims to enable breeding organisations to become more transparent.

Animal Welfare : In twenty years, most livestock production (pigs, poultry, dairy cattle, beef cattle) will probably be in large-scale units, though still mainly family owned. Animal welfare in this situation will be a high priority and animal breeding can contribute to it. In this context, attention should be paid to developing efficient information management systems for health monitoring, health detection, etc. Similarly, very little is known of the welfare of the domestic aquaculture species today.

New Technologies

Globally, there is active competition for new technologies that may directly or indirectly affect the future of animal production, including breeding. Some non-European countries (e.g. New Zealand, the USA, Argentina, Brazil, China) are rapidly developing research on ??and in some cases implementation of ? new reproduction and cloning technologies (somatic cell nuclear transfer, i.e. "Dolly-type" cloning) and genetically modified animals.

Other foreseen applications of the new biotechnologies in animals are in the medical field: animal models, animals as bioreactors, and animals for xenotransplantation. Although the development and possible use of such applications are beyond the direct scope of breeding organisations, Europe must be in a position to objectively evaluate these technologies and consider their potential.

Transparency and Dialogue

Transparency is of the essence, both in breeding and in the development of new technologies and pathways. Also essential is a continuous dialogue with the public as technologies are being developed. We need to discuss how increasing knowledge and the exponentially growing power of computer analyses can be applied in order to achieve balanced breeding effectively. We need to tackle questions such as: can finding the right balance enable us to meet global challenges? Can genomics be a powerful tool for this purpose? How can breeding knowledge and the organisation of breeding be exploited in order to satisfy the growing demand for animal products and to meet specific needs in relation to emerging economies, food safety, food supply, and the maintenance of diversity in small breeds?

Organisation of Farm Animal Breeding

Farm animal breeding is organised at different levels, from farms to breeding companies to breeding organisations up to the level of regulatory bodies providing national and international (e.g. EU) regulations. The organisation of farm animal breeding differs significantly according to the species. In poultry and fish farming, there are relatively few breeding organisations worldwide, as breeding work is knowledge intensive and relatively expensive.

In the case of ruminants, farmers are often organized co-operatively for technical services, representation activities, and breeding. Pig breeding is in an intermediate situation. In recent decades, only collaborative or specialized organisations have been able to breed livestock effectively, taking into account the latest scientific developments, using powerful computing systems and sophisticated estimation programmes, maintaining huge pedigrees of animals with their performances and biodiversity, and exploiting reproductive technologies. Yet there are no very big players in animal breeding. The organisations involved are (small or) medium enterprises or small units in larger organisations.

Poultry and Aquaculture Species

It is possible to concentrate breeding of small farm animals because the number of offspring per parent is high. This is what has happened. Margins being extremely low for eggs and meat from these species, breeding costs have been reduced as much as possible. In many cases former breeders have become multipliers (a step in the dissemination of breeding stock), outsourcing the selection work to another company. In poultry, for layer and broiler chickens and turkeys, 2-3 organisations are responsible for over 90% of the global breeding stock in each market. Aquaculture, a young sector, is developing in the same direction. Here, the number of species that can be farmed is rapidly increasing. Selective breeding is being applied to salmon, trout, seabass, seabream, and turbot, as well as other aquatic species such as shrimp and oysters. Breeding programmes for recently domesticated cod strains are now beginning. It is necessary to develop breeding and reproduction of additional species in aquaculture, so that closed selective breeding schemes can be initiated and the over fishing of wild stock can be avoided when the demand for aquaculture products increases.

Ruminants and Pigs

In the breeding of ruminants (cattle, goats, sheep) and still, to some extent, pigs, the animals' limited reproductive output means that a large number of breeding units (on farms) are needed in order to disseminate desired characteristics from elite animals at the top of the breeding pyramid. In addition, cows on dairy farms are the mothers of potential future breeding animals and are hence also a part of the breeding chain. The breeding of small ruminants such as sheep and goats is also specialised. The farmers of these species often live in marginal areas, far from the market and with a more limited access to technology than in the case of other species. Small-ruminant farming is often organised in a weaker co-operative system than for cattle. The importance of these species is likely to increase in the future because of their important environmental impact. New technologies in breeding and reproduction have not yet had much impact and should therefore be stressed.

In pigs, many farms are involved in disseminating top breeding animals to the farmer who grows the pig. Breeding programmes are often based on crossbreeding schemes integrating a dissemination level and a core breeding level. In ruminants and pigs, most breeding organizations are cooperatively owned by farmers. The organizations based in Europe are world leaders for their species. They are often SMEs operating in their native language, organised in national umbrella organisations. For disseminating desired genotypes, the national or local organisations of ruminant farm owners are internationally co-ordinated by a single international institution, based in Europe, created to provide services, mainly in animal breeding, on a larger, international scale. The largest pig breeding organisations in this sector are private or cooperative European organisations with a major impact worldwide.

Ownership

The purchase price of an animal includes its breeding rights, so the reproduction rights for an animal belong to the animal's owner. There is no animal equivalent of plant breeders' rights. The population structure of breeding makes such protective systems very sensitive to 'cheating'. Private investment in research, on the other hand, should be stimulated by dedicated protection of relevant findings.

It is important to strike a good balance between knowledge sharing and the protection of newly developed knowledge and tools. Patents are not commonly used in breeding – existing methodologies and knowledge should not be subject to 'broad claims' and 'new claims' that would hamper access to methods already applied or described and discussed openly at round tables and conferences, in scientific magazines, etc.

Regulatory Aspects

The organisation of animal breeding and reproduction is under permanent supervision by ?and interaction with ??regulatory bodies providing national and international regulations (e.g. EU authorities and ICAR, the International Committee for Animal Recording). Interaction with these bodies is crucial to promoting the harmonious development of animal breeding and reproduction practices, especially in ruminant farming with its open breeding structure.

A Vision for Sustainable Breeding and Reproduction

Animal breeding and reproduction have a great contribution to make to a future sustainable animal agriculture. Many opportunities are open to the animal breeding and reproduction sector for improving the biological and economic efficiency of food production and increasing food supply. These opportunities are particularly attractive in a sustainability perspective.

The sector is intimately linked to all three pillars of sustainability: Consumer, Environment, and Economy, through its interactions with societal developments, its implications for biodiversity and the environment, and its contribution to economic growth. How these three pillars will develop in the next twenty years and how sustainable breeding in the EU will fit into the picture are questions that must be addressed.

Sustainable breeding and reproduction means balancing:

- safe and healthy food
- robust, adapted, healthy animals
- biodiversity
- social responsibility
- a competitive and distinctive Europe and must include new prospects for animal production in Europe. When it comes to developing new technologies, transparency about the developments and an open dialogue with society are prerequisites.

Integration into Animal Agriculture

Traditionally, animal agriculture and aquaculture are fragmented into distinct activities and focuses: animal nutrition, animal waste management, food science applied to animal products, animal welfare, animal management, and animal breeding and reproduction.

Given the challenges facing animal agriculture and aquaculture today, we believe that a much higher degree of integration is needed. In our vision, integration will affect all levels and will certainly influence the future of animal breeding and reproduction.

Safe and Healthy Food

Product Quality

In a "fork to farm" production system driven by consumer needs, quality in the broadest sense (food safety, nutritional value, sensory and technological quality) becomes one of the most important drivers. Safe and wholesome food is clearly the first requirement. Beyond safety and for all animal products, product quality improvement will be a major issue. This is a significant shift from the food sector's position in the last century, when improving the level of production was the obvious goal of a supplyoriented production system.

In the modern context, nutritional value and human health features (e.g. the fatty acid composition of meat and milk, the nutritional quality of eggs) can become targets for animal breeders, as can sensory qualities such as tenderness, flavour (importantly, this includes boar taint), visual appeal, and processing characteristics. The importance of breeding systems and new technologies in protecting the quality and enhancing the economic efficiency of local and typical productions deserves full consideration.

Safe Gene Dissemination

The safe dissemination of animal genes is a major issue. The risk of transmitting diseases through animal, semen, egg, and embryo transport has increased with increasing farm size and the internationalisation of trade. It is essential to minimize this risk. Artificial Insemination and embryo transfer have contributed a great deal towards this goal, but there is room for further reducing the disease transmission risk and ensuring safe transport of genetic material by improving the health guarantees of this material. The role that existing and new reproduction technologies can play in the (international) transport of breeding material should thus be weighed against the ethical dimension of using certain technologies (e.g. embryo technologies). Specific Pathogen Free (SPF) production of breeding animals results in animals free of certain diseases, which in turn leads to improved human and animal welfare.

Robust, Adapted, Healthy Animals

Animal Welfare

Breeding organisations ensure the health and welfare of the animals they keep and select. They are engaged in the search for selectable traits that are indicative of species-specific animal welfare. New biological insights into brain function, the genetics of behaviour, and physiological indicators of stress and well-being will provide new tools enabling breeders to handle welfare traits more objectively than at present.

Adaptability

Breeders produce animals for a wide range of production environments, from extensive and/or organic systems to more intensive systems on larger farms (which continue to get even larger). Breeders want to be able to improve the level and efficiency of production in each of these environments. It is essential that animals remain healthy and productive under this wide range of environments and with reduced human interference (labour costs are the main costs in farm animal production, along with feed). It is now common for breeders to test animal performance in more than one environment, but new knowledge on how animals adapt will enable breeders to take novel aspects of production into account more effectively in breeding programmes. Over the next twenty years there will be more large-scale units and new technologies for which an efficient information management system will be developed. Aims will be to facilitate selection of animals for adaptation to these systems and to use technology to support balanced breeding programmes.

Domestication of Fish

Most farm animals (and companion and sporting animals) were domesticated thousands of years ago. This is not the case of aquaculture species. Here, the domestication process has just started. A lot of work has to be done towards improving the reproduction of aquaculture species and towards domesticating species that are now only caught in the wild.

Disease Resistance (General and Specific)

Breeding may contribute to robust and healthy animals by selection for more general and specific disease resistance as is done for aquaculture species today. This should result in less use of medicines. Optimising the use of medicines and vaccines according to an animal's genetic makeup represents a significant development opportunity.

Balanced Breeding and Biodiversity

Balance

It is necessary to further increase production levels so as to lower cost price and to ensure an adequate supply of meat, fish, eggs, and milk for a fast-growing human population demanding more of these products. Yet especially in Europe, breeding cannot be driven by production efficiency alone. In defining breeding objectives, it is necessary to strike a good balance between the production level, animal physiology (welfare, behaviour, health, reproduction), and population fitness. By learning more about animal genes and physiology and by exploiting greater computing power, it should be easier to balance these multiple breeding objectives. To achieve this optimally, we need more knowledge and better tools in areas such as the basic biology of traits and their interrelationships and data handling. Also essential is the improvement of genetic and reproductive tools with regard to precision, ease of use, and cost.

Biodiversity

Breeding programmes are designed to make optimal use of existing genetic variation between and within populations. Breeding organisations must contribute to maintaining genetic diversity in their breeding populations. They must monitor and control the rate of inbreeding and genetic drift. In animal populations with an open structure (e.g. cattle), the 'effective' population size is much smaller than the actual population size. A research opportunity is to better understand the interaction of biodiversity and genetic variation. This will foster knowledge-building in diverse areas, such as the co-evolution of hosts and pathogens and animal adaptation to climatic differences or variable nutrient availability. It will also contribute to defining appropriate levels of biodiversity, identifying in small populations 'rare' gene variants whose preservation should be prioritised, and protecting genetic diversity in populations used for large-scale production.

In the coming years, more attention will have to be paid to managing (as distinct from conserving) animal genetic resources, particularly given the sometimes narrow genetic base of pigs, poultry, and dairy cattle, and the concentration of breeding strategies and selection in very few hands. Furthermore, it is important to bear in mind that biodiversity preservation is the key to preserving future breeding opportunities. It is essential to maintain sufficiently diverse gene pools because we cannot foretell the changes that might affect the social and economic driving forces of animal agriculture. Nor can we predict the evolution of biodiversity-related ethno-zootechnical issues that influence the link between farm animals and society. The preservation of biodiversity is also important for conserving and improving local breeds, which are very significant in marginal areas. The need is to protect these breeds by increasing their production and standardising the quality of their products. Objectives will be to defend the environment of marginal areas and to maintain the efficiency of animal farming in less favourable areas.

Social Responsibility

Environment

There is scope for selecting animals better suited for maintaining biodiversity under changing climate conditions. A major achievement of animal breeding in recent years, and one with a highly favourable effect on the environment, is the improvement of feed efficiency. When animals make more efficient use of nitrogen and phosphate, this minimises the output of these elements into the environment. The main contributor to improved feed efficiency is improved feed conversion. Yet there is room for further improvement of this process, especially in farmed fish.

Progress will require new knowledge on the genes and pathways that may contribute to improving the European environment while an animal's production level is increasing. Opportunities for feed efficiency improvement exist in all farm animals, from fish to cattle. This requires better knowledge of the digestive system and of nutrient metabolism. Of particular interest are the genes whose expression is regulated by nutrition. When exploiting such knowledge in animal selection, it is necessary to take into account animal product quality and both beneficial and unfavourable environmental outputs. Animal breeding contributes to the environment in yet another way: through selection of animals well adapted to their environment and interacting favourably with it by grazing and pasturing. Such animals contribute significantly to ecosystem maintenance. The selection of genotypes for the maintenance of the environment is fundamental in many European geographical areas, especially mountain and Mediterranean areas.

Animal Welfare: Integrity

Animal integrity is defined as "the wholeness and completeness of the species-specific balance of the creature, as well as the animal's capacity to maintain itself independently in an environment suitable to the species." Most farm animals have been domesticated for thousands of years and as such have been adapted to an environment of co-existence with humans. They are easier to handle and able to reproduce in captivity, under conditions where food is abundant and there are no direct predators. Breeders must keep a balance between the intrinsic characteristics of domesticated species, their welfare, and improved production.

(Contributing to) World Development

The global demand for animal-derived foods is increasing. European breeding organisations play a major role in providing breeding stock globally. For pigs, aquaculture species and poultry, which are reared in relatively controlled environments, this has generally been a huge success. The history of cattle breeding, however, is littered with mistakes. Highproductivity European breeds are not always suited to more extensive environments characterised by a different and sporadic food supply and by various climatic and disease challenges. It is possible to learn from these mistakes. In global development programmes, the knowledge and organisational skills of European breeders could be adapted to meet local needs in developing countries.

New breeding schemes require multiple-character selection based on both production traits and other important traits such as resistance to disease (sanitary quality and animal welfare), fertility, and longevity (animal welfare). At the same time, there is a need to conserve biodiversity through the development of appropriate breeding strategies in rare ruminant breeds.

A Competitive Europe

A Strong Breeding Sector

A prerequisite to breeding in Europe is the survival of the breeding sector. Major companies have premises for breeding and reproduction outside Europe, where research can often be more easily funded and carried out. There is a constant push to move away from the head offices in Europe in order not to lose market potential and ultimately lose out in the competition. Every year breeding companies disappear or amalgamate. When companies move away, they will not come back. An important challenge and opportunity is to keep the breeding climate in Europe challenging and interesting enough to maintain businesses here.

Breeding needs profitability at all levels and requires active participation of farmers in order to test animals for desired characteristics and to disseminate those characteristics. This is feasible only if such endeavours are economically sustainable. It is crucial to preserve the livestock system by maintaining and/or increasing the efficiency of breeding also in small-scale animal farming. Many of the new EU member countries have a tradition of small-scale aquaculture, which needs to become more efficient in order to compete. Breeding is a global activity, and European businesses have gained leadership in the breeding of several farm species. Yet the market remains competitive and as technology developments continue at a breathtaking pace, European breeders must not be complacent. There are structural weaknesses in Europe compared to other global players, as breeding is often organized nationally. Given the existence of 25 different countries in the current European Union, this is a natural disadvantage for European animal breeding.

Further strategic partnerships in research, development, and business across Europe will be increasingly important. Research opportunities will arise in operational genetics, numerical biology, animal recording, data management, and highthroughput biological research. If Europe is to retain its competitive strength, there must be sufficient support from research funders to ensure that the basic and strategic research opportunities are seized.

European public funding should play an important role in this context, helping to foster integration of research activities and co-ordination of research and innovation policies. It is essential to bear in mind that the changing world will demand careful but fast adaptation of farm animal production systems and hence of farm animal research and breeding in its broadest context.

Transparent Public Engagement

New technologies, goals, or practices will be part of the road to the future. In some parts of the world people are already applying food production technologies that Europe considers unacceptable. This trend towards parallel development paths may well continue, so Europe must develop alternative approaches if its agriculture is to avoid facing even greater competition than it does today from cheaper imported foods. One promising pathway might be a trend towards differentiated foods (e.g., speciality foods, organic foods), as food production in Europe is likely to develop towards quality instead of quantity.

Alongside the development of new technical solutions to breeding problems, it is important to involve specialists in ethics to support careful interaction with society as a whole. It is vital to realise and take into account which breeding activities are in line with what European citizens want animals to do and produce and what they consider inappropriate. Transparency about breeding work and research developments is a key issue for farm animal breeding scientists and breeding organisations. Animal breeding strategy over the next twenty years must include a commitment to communicate with civil society. In the past, a lack of information has sometimes given new breeding technologies a negative public image.

A Flexible, Open Environment

For European breeding organisations, competition at the global level is strong. Unlike breeders in the largest competitor country, the USA, European breeders must deal with a multiplicity of scientific, funding, and regulatory environments within Europe. European breeders need a flexible, open environment in order to continue their business and to develop and implement the schemes needed for the future. They are therefore in favour of initiatives such as regular and timely consultations with regulatory agencies, dedicated to finding workable yet safe solutions for animal transport and new technology implementation (guided by ethical and safety considerations), investigating specific conditions for minor species or species under development, and creating the same conditions for imported material as for European material.

Well-Educated People

For a knowledge-intensive sector, the education of students and hence of future science and industry leaders is important. The growing complexity of animal breeding makes this especially true. Today the sector requires a knowledge base combining largely quantitative science with molecular genetics and genomics, as it pogresses towards the day when it will be an industrial user of predictive biology. Education, in the future, should be a more collaborative endeavour where public and private bodies cooperate more closely.

A Distinctive Europe

Rural Areas

Farm animals, especially ruminants but also horses, have an important role in maintaining the European rural landscape, e.g. in keeping areas clear to prevent fires. Only a few local breeds are now still part of that landscape and add to its diversity across Europe. Therefore, we need to maintain those breeds that fit with the landscape and, in harmony with the environment, yield high-quality local products. We need to breed them to maintain the landscape whilst improving their economic sustainability.

To reach these objectives, it will be necessary to consider breeding goals and technologies for extensive systems. The value of the resulting breeding system will be enhanced by the enormous importance of extensive livestock farming not only for production but also for its environmental and social benefits. Aquaculture production has another value for the rural areas as it is often situated in such areas and can therefore assure that work-places are secured there.

Regional Products

Europe is a diverse continent with typical products related to local breeds, local environment or climate, or local habits or history. Consumers often go for a value-for-money product, but increasingly they appreciate the perceived added value of regional products. In many situations in Southern Europe and mountain areas, breeding will have to be planned to improve the distinctive qualities and local characteristics of animals. This will lead to a differentiation of food products.

Social/Cultural Aspects

European animal production is also important for the vitality of the countryside. When farmers disappear and holiday homes take over, the social infrastructure can suffer (e.g. schools, shops, public transport). Cultural values may open a market for some higher value niche animal products, appreciated by consumers when the origin of the food is part of its attraction (e.g. when eating for pleasure in a restaurant or cooking a special meal).

A Diversity of Benefits

Animals for many Purposes

With the standard of living increasing and more time available for most people, animals for pleasure and leisure are a growing industry. In addition to companion and sporting animals, there is a growing trend towards 'hobby' farming. Animals will also be bred for social purposes, e.g. for 'animal therapy' or for helping to rescue people in an emergency. Breeding for the right type/character of animal for the right purpose may be a new, promising prospect for breeders.

Fibre and Skin

Wool, leather, and skin are among the traditional farm animal products. They used to be necessary items for clothing, but nowadays they are used for their 'naturalness' (e.g. wool) or for the specific advantages of the product (e.g. leather). Breeding sheep or other fibre animals for typical products, like natural coloured wool, is an increasing niche market.

Animal Biotechnology

The combined advances in genetics, embryology, and stem cell research open the prospect that high-addedvalue products for agricultural, medical, and technical applications will soon become a reality. The major applications will be in the use of highly specialized animals for drug production (e.g. in milk), for xenotransplantation, or as animal models in research on human diseases.

In the future, animals may become precious tools for modelling human disease and developing new therapeutic strategies. Such approaches will both complement and inform studies based on human cell/tissue cultures.

To make such applications possible, it will be necessary to develop robust embryonic stem cell lines capable of self-renewal in culture and to refine the technologies used for cloning and transgenesis. These same technologies will be valuable research tools enabling scientists to gain deeper understanding of livestock and companionanimal embryology and reproduction.

A challenge in the next two decades will be to integrate and potentially exploit these novel technologies in a society-friendly manner. For this it is essential to build the knowledge base and expertise required to provide technological guidance, quality control, and safety.

An Agenda for Research and Technology

Quantitative Genetics and Operational Genetics

Some inherited traits, such as milk yield or daily gain, are controlled by many genes. In an animal population, different animals will have different combinations of gene variants affecting a given trait. This leads to quantitative variation of the trait within the population. Such traits and the underlying genes are the focus of quantitative genetics. Over the past fifty years, most improvement in the quality and efficiency of farm animal production has been due to the application of quantitative genetic methods. For instance, the International Bull Evaluation Service (Interbull) maintains and provides access to a huge body of genetic animal evaluations from the most advanced countries. Its ranking of bulls according to their 'genetic merit' facilitates the genetic market worldwide. Operational genetics is the application of genetics to breeding, for example in designing breeding programmes and optimising selection schemes to ensure maximum genetic gain while minimizing inbreeding or undesirable trade-offs of selection. If breeders are to maintain and build upon the enormous genetic progress made in past decades, they need the support of research focusing on the further development of complex statistical methods and computer algorithms for coping with multiple traits. They need information from new sources such as genomics and the study of how animals adapt to specific environments. Attention must also focus on the more complex, non-linear, and difficult-tomeasure traits such as disease resistance, longevity, and robustness, as well as on new and emerging traits. Alternative selection schemes should also be envisaged for the analysis of non-additive genetic variation.

Advanced Modelling for Management Purposes

Quantitative genetics has a favourable 'side effect' of which people are not always aware: extensive modelling of complex phenotypes (biological traits) provides interesting and important information about environmental influences on these phenotypes and makes it possible to predict unobserved phenotypes. It will be important to incorporate this and all relevant knowledge into advanced management tools, which will notably increase the return on investment of extensive data recording schemes. Current efforts in the field of milk recording (test-day models) are just the start of a whole new range of research opportunities. Future research in statistics and computing sciences will be required to optimize current methods and procedures. As an on-farm management system develops, a challenge will be to link it to more integrated, multi-level systems.

Phenomics

Phenomics means the measurement of animal phenotypes. Currently animal breeding programmes can only improve measurable traits or traits genetically related to measurable traits. They rely on characterising animals for production, quality, and welfare traits on the basis of limited numbers of laboriously collected records. The entire selection system for ruminants, through which many traits have been improved, is based on this traditional approach, embodied in a manual used throughout Europe and on other continents: the 'ICAR Guidelines for Animal Recording'. In the future such methods are likely to be complemented by high-throughput automated recording systems, on farms or in processing factories for example.

An important future research focus will be the estimation of carcass meat content by means of automated image collection. Accurate measurement of many other meat traits (e.g. colour, fillet yield) currently requires animal slaughter, but new techniques have emerged (Near Infrared Spectroscopy (NIR), ultrasound and computer tomography) that will allow such traits to be recorded on the breeding animal. This will increase the accuracy of selection and hence, the genetic gain. Ultrasound techniques were successfully implemented in pig and beef cattle breeding in the 1970s and can be further developed in other species. In dairy animals, more sophisticated techniques for the recording of milk yield and composition are necessary in order to continue improving the production efficiency and nutritional quality of milk and milk products through selection. Much remains to be learned about how biological attributes and genetic features influence many sensory traits of meat. For example, are variations in meat texture due to muscle fibre number, fat content, the amount of connective tissue, or to all of these factors? Before breeding programmes can tackle such traits, we need to know which measurements are needed to describe them and how to interpret those measurements.

Genomics and Beyond

An organism's Genome is its Genetic Material (DNA)

Animal genomics, broadly defined, is the study of animal DNA, its organisation into genes, the roles and interactions of gene products (mostly proteins), the control of gene expression, and ultimately all downstream impacts on animals, their traits, and their interactions with the environment. The sequencing of the human genome was a milestone in the development of modern biotechnology, opening new avenues for understanding the control of complex traits at a molecular level. The genomes of several farm animal species (cattle, pig, rabbit, and aquacultural species) are also being sequenced and analysed. This will facilitate the integrated analysis of biological functions.

At sequence level, genomic research is yielding genetic markers for selection, pedigree control, and traceability. It is facilitating the identification of trait genes for use in breeding programmes. Beyond this level opens the whole new realm of "omics" research: transcriptomics (focus: expressed genes), proteomics (proteins and their functions and interactions), and metabolomics (metabolites and metabolic pathways). This research is bridging (part of) the gap between 'structural' genomic knowledge (markers, maps, sequences, genes) and everything we hope to learn by exploiting it. It provides a bridge to nutrition, health, and more. It is yielding precious knowledge on a wide range of processes related to health and food safety, such as acquired resistance to contamination through gut health or the long-term effects of an altered maternal diet (foetal programming) in both animals and humans. The application of such knowledge (in marker-assisted breeding programmes, for instance) will contribute to improving animal health and food safety, animal welfare, and the biodiversity of breeding populations. In the long run it will boost the competitiveness of European farm animal breeding.

Functional Genomics

This area includes transcriptomics and proteomics. It aims to elucidate gene functions. Transcriptomics uses powerful tools (e.g. microarrays: arrays of material resulting from gene transcription, the first step in gene expression) to study simultaneously the expression of many (ultimately all) genes in a genome in various situations (specific developmental stages, environmental stress...). Proteomics exploits various technologies to study which proteins are present in ? or secreted by ??different cells under different conditions, and also looks at protein interactions.

Functional genomics is used to study genes that contribute jointly to a specific function as part of the 'interactome' (the whole set of molecular interactions in cells). This notably includes networks of interacting proteins, structural and mitochondrial DNA, and inhibitor RNAs (RNA is quite similar chemically to DNA; it plays multiple roles, notably in gene expression and its regulation). By combining multiple approaches and tools, it should eventually be possible to build models linking together large numbers of genes and their products, in order to understand the impact of gene variation on animal traits, mediated through the relevant metabolic pathways. To achieve this, a considerable research effort is needed, yielding better understanding of animal biology on the basis of knowledge obtained from the genomes of model organisms and humans.

Exploring Within-Species Variability

To allow effective selection of animals via genomic approaches, it is necessary to explore within-species variability and to quantify factors such as linkage disequilibrium (the tendency of genetic features that are close to each other on the genome to be inherited together, and thus frequently to be found together in the animals composing a population). Effective genomic selection requires the development of tools for genotyping animals by marker typing or direct sequencing. It requires improved tools for phenotyping animals more exhaustively and quickly so as to identify genes that control variation of multiple traits. We also need to develop strategies for predicting phenotypes on the basis of genotypes. Here research should notably focus on the genetic control of variation in gene expression and the identification of genome regions undergoing epigenetic modifications such as imprinting. Given the strong negative impact of inbreeding on reproductive and health traits, we need to find ways to reduce the inbreeding coefficient in pigs, poultry, aquaculture species, and dairy cattle.

Exploring Between-Species Variation

By learning more about between-species variation, it will become clear which vital processes program which functions in animals. This will enable breeders to identify animals yielding healthy animal products without this being detrimental to their own health. With genome-scale sequencing of domestic species, between-species comparisons (including comparisons with mouse and human sequences) will make it possible to identify variability 'hot spots' in genes and/or genome regions so as to focus the search for single nucleotide polymorphisms (SNPs) responsible for substitutions in protein sequences in domestic animals (SNPs are sites where a single 'letter' in the 'text' of a gene is variable).

Biology of Systems and Traits

Currently the basic biology (genetics, biochemistry, and physiology) underlying the variability of most traits of interest to animal breeders is poorly understood. This is an obstacle to controlling physiological processes that influence such traits. Molecular genetics will yield abundant novel information on the fundamental regulation of such processes. Here research must aim to provide detailed operational understanding of the pathways involved.

Many gene products influence several metabolic pathways, and most metabolic pathways are influenced by several gene products. This makes for very complicated physiological network patterns. In addition, whole sets of genes can be switched on or off via so-called 'epigenetic' mechanisms (DNA methylation, histone acetylation, RNA interference) participating in complex regulatory networks. Such mechanisms are increasingly found to play significant roles in phenotype development and transmission. Control of a single gene may thus cause unwanted side effects when the physiological network around it is not sufficiently understood. Future research must focus on dissecting the genetic basis of traits of relevance to sustainable animal breeding and the Inter-relationships of these traits.

Whole-Animal Biology and Gene-by-Environment Interactions (GxE)

Understanding individual traits is not enough. What matters is the biology of the whole animal and how an animal's genes interact with its living environment to determine its quality of life and performance. Biological research is already moving towards predictive biology. A first step is to model the various systems that constitute the biology of a cell (creating an e-cell) so as to predict cell responses to environmental change (e.g. altered nutrient supply). The next step is to build models that describe cell-cell interactions in tissues and finally to build upward from tissues to a whole animal. This is a considerable challenge, but it is undeniably the direction in which biology is heading. High-throughput laboratory methods now make it possible to measure many components of a system in parallel, and these approaches complement the traditional whole-animal research supporting farm animal production. With modern research techniques, scientists will be able to revisit 100 years of animal science with a view to expanding knowledge of all biological mechanisms. Future research must focus on developing an integrated understanding of the lifetime functioning of whole animals and their interactions with the environment. At each step it will be essential to evaluate progress in the light of breeding reality. What breeders need especially is a detailed understanding of relationships between performance, health, welfare, and environmental impact combined with costeffective, robust, reliable ways to measure traits reflecting these relationships in animal populations.

Population Biology

Future qualitative and quantitative genetic technology will give breeders much control over the genetic makeup of individual animals selected for breeding. In other words, what can be selected with increasing precision is the genome of an individual. Yet population-level effects also deserve consideration for at least three reasons. Firstly, social interactions in groups of animals are associated with behavioural repertoires that are important for animal welfare and for proper functioning of the group: maternal behaviour, dominance and aggression... To the extent that such interactions are linked to production traits, selection could have negative side effects when the physiology of social behaviour is not sufficiently understood. Secondly, the production potential inherent in an animal's genes is variably expressed according to the nutritional, climatic, infectious, and social environment.

This is called environmental sensitivity or phenotypic plasticity. It provides a measure of genotype-environment interactions at the level of an individual animal. Future genetic technology will offer increasing control over environmental sensitivity, and again side effects are likely if interaction mechanisms are not properly understood at population level. It is important to take into account the dynamics of the environment in which future breeding animals will have to produce. The environment will change in the next twenty years and strategies for the genetic improvement of animal populations must adapt to these changes. Also important is the development of areas such as the improved management of pastures and silvo-pastoral systems for increased overall efficiency in animal production. Research should focus on defining, through population biology, which will be the most efficient genotypes for predictable environments.

Thirdly, disease transmission is perhaps the most crucial population-level issue. Infectious disease involves two (or more) 'actors': host (the farm animal) and pathogen. These two players interact through infection (pathogen to host) and immunity (host to pathogen), and the pathways involved in both processes are under genetic control. Future genetic technology will provide much knowledge about regulation on each side of such a system, but the connection between the two, i.e. the host-pathogen interaction, is likely to be the crucial element in getting the system under control. Exploring this connection means focusing on epidemiology in relation to pathogen virulence and host resistance. It requires studying the co-evolution of hosts and pathogens. Research exploiting emerging technologies is bound to provide much knowledge on the regulation of the immune system at the individual animal level, but turning this knowledge into operational control will require linking it properly to population-level epidemiology.

Future research must focus on understanding how variation in a population affects the overall performance of groups of animals and how changes in biodiversity are likely to influence the control of infectious disease. This will ultimately lead to defining appropriate levels of biodiversity for the long-term sustainability of animal agriculture and aquaculture.

Reproduction Technology

We need research on the reproduction technologies required to underpin breeding and the effective dissemination of genetic improvement to all producers. Artificial insemination is long established in many livestock systems as a central method of animal reproduction with an essential role in breeding programmes and genetic dissemination. More recently, additional techniques have been developed to further improve genetic progress, animal health, and animal welfare, and to disseminate breed improvement. Technologies such as embryo transfer, cloning, and sexing are under development and have had some limited application. Initially, future work on reproduction technologies will aim for a better technical, physiological, and genetic understanding of current and new technologies such as nuclear transfer and cloning for research purposes and semen control of sex determination (e.g. through semen sexing).

This will include developing more reliable predictors of semen and embryo quality and studies into the interactions with the environment, nutrition, and infection. Benefits should also stem from research on more reliable indicators of oestrus, conception, and the survival of cryopreserved gametes and embryos. In aquaculture, additional techniques (e.g. for sex determination) and knowledge on the limitations of reproduction are needed. Future research on cell differentiation may open the way to producing gametes from stem cells. Coupled with predictive biology and statistical techniques such as genome-wide selection, these approaches could make it possible to produce and select multiple generations in the Petri dish. Clearly this entails significant risks as well as potential advantages, so such developments are probably beyond the timescale of our vision.

Reproductive technology research will contribute to safe and wide dissemination of valuable genetic material so as to achieve genetic progress and quick transfer to the production level. Transparency about new technological developments and an open dialogue with society are important in such a sensitive area. In some livestock systems, such as beef cattle, sheep, goats, horses, buffalo, and most aquaculture species artificial insemination (AI) is still very rare. Reasons for this are many. One of the main reasons is the difficult extensive environment in which these species are usually bred, while AI requires a minimum level of technological input. This applies to a rather large part of livestock production in Europe, and therefore adaptation of the required technology is certainly important.

Biotechnologies

The new cell-based strategies described above have the potential to add to the desired developments in farm animal breeding. More controversial technologies such as genetic modification could be even more powerful, delivering step changes in sustainability in ways unlikely to be possible through conventional selection. For example, there are GM pigs in Canada claimed to have a reduced environmental impact because they have been given a digestive enzyme not naturally present in pigs. This enzyme allows better utilisation of phosphate in the pig's diet, reducing the pollution per unit output. It is debatable whether these pigs will ever be economically viable, but they illustrate the current state of technology. Future research is likely to deliver other GM options, one of the most exciting being the potential to make animals completely resistant to specific viral diseases. If genetic modification (RNA interference) could make chickens resistant to avian influenza and reduce the risk of a human flu pandemic – potentially saving millions of lives – would this be an acceptable use of genetic modification?

Despite current public sensitivity, research in the US and Asia is likely to increase pressure on EU agriculture to accommodate novel methods, including the use of nuclear transfer ("cloned") animals for breeding in the pig industry. It is noteworthy in this regard that the US Food and Drug Administration is expected soon to allow products from cloned animals to enter the food chain, with potentially significant implications for international trade, via WTO negotiations for instance. Research in these areas can be part of a medium-and long-term strategy. It can provide information on risks and benefits, enable European society to make informed decisions, and help breeders to define future economic strategies. Research and appropriate communication can also accommodate potential shifts in European public perception in the coming decade.

Genetic modification, research on large-animal embryonic stem cells, and novel methods we have not yet thought of will provide valuable research tools for reducing the number of animals used in research (why do the experiment on an animal if it can be done on a specialised cell line in tissue culture?). Plant breeders are already 'stacking' multiple genetic modifications in an individual crop. Animal breeding is some years behind, but targeted gene modifications in large animals are required for controlled experimental work. Future research should thus include a focus on: i) basic research in animal embryology and reproduction, ii) deriving farm animal stem cell lines and tools for controlling the differentiation of these lines, and iii) cloning and gene transfer methods. The need for this type of research stems from Europe's wish to be a knowledge-based economy while taking into account the ethical climate in Europe. It does not stem from current application requests in the animal breeding industry. Yet to adopt a science based stance, Europe must support the development of new technologies within its borders. We thus strongly support European and national research into new technologies, also in farm animals, as basic research, to be conducted in a transparent way. This includes dialogue with stakeholders and society.

Socio-Economic Research

Society and societal attitudes are constantly evolving and are crucial to defining the ethical, legal, and market context in which animals are produced and animal breeding decisions made. Worldwide and even across the EU-25, there is considerable variation among individuals in their attitudes and behaviours towards animal agriculture and aquaculture. It is consequently very difficult for animal breeders to predict future trends. If animal breeding research is to be relevant and animal selection decisions are to be appropriate, biological research must go hand in hand with socio-economic research shedding light on issues such as segmentation among consumers and how beliefs and behaviours interact. What are the range and distribution of consumer attitudes towards foods of animal origin? How do consumers weigh different social considerations (animal welfare, traditional practices, regional distinctiveness...) versus price in their purchase decisions over time? By answering questions such as these, socio-economic research may notably help build models of consumer attitudes towards novel technologies.

Here are some of the topics that socio-economic research should address: the role of animal agriculture in sustainable rural economies and rural environments (for this a holistic (life-cycle) approach is advisable); developing better tools to explore how best to legislate on biotechnology matters and how legislation will affect European competitiveness, consumer health, and food prices; developing effective strategies for communicating information on animal breeding science and technology and for transparency. Lastly, since developments in human genomics are likely to lead to personalised dietary advice on 'good' and 'bad' foods for individuals, an important question is: how will this affect the consumption of animal-derived foods and will this further increase the segmentation of the market?

Animal Science

Equipped with this vast and growing array of extremely powerful technologies, animal breeding is able to address a great many industrial and societal needs. Yet selective improvement of farm animals must be undertaken in a context of continuing general animal science. As the genetic makeup of farmed animals changes we can expect the nutritional needs of the stock to change as well. This will have an impact on which husbandry practices are most appropriate. Producers with the latest breeding stock also need information and advice on the best available management technologies. Therefore, research along the various lines described in this chapter must be complemented by parallel research in animal production science. Especially important will be redefining the nutritional needs of improved genotypes and adapting management systems to optimise the benefits of selective breeding. The interaction between management and breeding must be acknowledged and will be an important issue in the future.

Chapter 3

Available Technology and its Application

Tropical pasture technology aims at improving animal production through the judicious use of inputs according to the prevailing economic and social conditions effecting the production system. Forage production can be improved by establishing a) grass pastures requiring fertilization with NPK, b) grass-legume mixtures, c) or oversowing legumes into existing pastures, d) monocultures of legumes (protein banks) and e) fodder crops.

The use of fertilisers on legume based pastures would consist primarily of Phosphorus (P) and rarely Potassium (K) plus trace elements. For intensive production systems on poor soils, fertilization would materially increase both the yield and persistence of pastures and would provide a good return on investments.

Grasses are, however, more difficult to establish and are less persistent without fertiliser (particularly N) than legumes, although some legumes will not persist on poor soils without P and trace element supplementation. However, in many situations fertilization is not practical because of the low return from animal products, in such cases legumes such as *Stylosanthes* species can introduced which will produce even under low fertility. How the forage will be utilised (grazing or cut-and-carry) is also important.

With grazing nutrients are returned, albeit very unevenly, but this is rarely the case with cut-and-carry systems where the nutrients are removed along with the forage and dung (and urine) is used for other purposes (fuel or manure). Repeatedly cutting forages rapidly depletes soil fertility and fertilization is necessary to sustain production. Forage production is not dependent on pastures and forage crops alone and crop residues can make an important contribution in integrated crop/forage production systems. Two examples of the latter are the integration of livestock under tree crops (Reynolds, 1988) and alley farming. The most appropriate method of pasture improvement depends on the intensity and the objectives of the production system.

Dairy production for the liquid milk market, with assured income and high requirements for quality feed, can justify more intensive production methods than say beef production in remote regions which rely on uncertain markets for low quality beef. The dairy farmer is more likely to base production on intensive, fertilized grasses or grass/legume mixtures; whereas the beef producer might consider, if anything, oversowing legumes into native pastures without fertilisation.

Numerous selected grass and legume varieties have become available since the 1960's for all tropical and subtropical regions with the exception of the semi-arid and arid zones (Henzell and 't Mannetje, 1980; 't Mannetje, 1984). Appendix 1 lists the main grasses and legumes for different climatic zones as defined by Troll (1966):

- tropical rainy climates with 12 to 9.5 humid months (Zone V_1)
- tropical humid-summer climates with 9.5 to 7 humid months (V_2)
- wet and dry tropical climates with 7 to 4.5 humid months (V_3)
- permanently humid climates with hot summers and maximum rainfall in the summer (IV_7)
- climates with 9 to 6 humid summer months and dry winters (IV_4)

The main requirements of forage species are that they are adapted to the prevailing climatic and soil conditions, tolerant of grazing, disease and pest resistant, free of harmful constituents (non-toxic) and tolerant of low fertility. It is not possible to find species to which all these requirements apply, however, management can make up for species' shortcomings.

For example, legumes which have a low grazing tolerance could be replaced by those that do posses it or grazed intermittently. Another example is the mimosine content of *Leucaena leucocephala*, which need not be a problem to animal health so long as the animal's diet does not exceed 30 % of the diet for more than six months.

Alternatively, it is also possible to infuse the rumen of susceptible animals with micro-flora which can break down the toxic derivative of mimosine (DHP) (Jones and Metgarrity, 1986). Not all species need to be tolerant to low fertility because fertilizers can be used in certain production systems.

The Application of Pasture Technology

The decision to improve pastures in existing farm systems is governed by the following conditions:

- the need for improved animal production
- the value and marketability of animal products
- the motivation of farmers
- land tenure, and
- the availability of resources.

The Need for Improved Animal Production.

Animal proteins are an important component in the human diet and in most developing countries there is a need for both more and better quality food. Ruminants are able to produce food, especially proteins, from land which would otherwise be of little use for food production.

When comparing the distribution of people, farm animals and the production of meat and milk between temperate (mostly developed) and tropical (mostly developing) countries, it is clear that tropical countries lag behind. Although far more animals are kept in the tropics, the production of meat and milk is only 36% and 18%, respectively, of the total world production.

This is due to inadequate nutrition, health care and management of the animals. Therefore, if animal production is to be increased, not only for meat and milk, but also for draught purposes and manure, then the feed supply must be improved. Since forages and crop residues are the primary feed sources, it is evident that both production and quality of forage needs to be increased.

An increasing demand for forages will also result from a) the increase in the demand for food as human populations continue to expand, and b) increasing purchasing power which leads to a greater propensity to buy animal products.

The need for pasture improvement can be shown by the low levels of animal production in both developed and underdeveloped tropical regions. This will be illustrated by examples from northern Australia and Africa.

Table: *Distribution of farm animals, people, production of meat and milk (source FAO Yearbooks)*

	% in developed countries	***% in developing countries***
Cattle	36	64
Buffaloes	—	100
Sheep	49	51
Goats	6	94
Pigs	40	60
People	25	75
Milk production	82	18
Meat production	64	36
Kg milk/person	320	23
Kg meat/person	54	11

Northern Australia: In northern Australia with its sub-tropical to tropical climate, large scale beef ranching is practised on unimproved native pastures. In the Northern Territory the average property size is 2500 km^2, the stocking rate is 6–8 beef animals per km^2 and the main emphasis is on breeding. Branding (weaning) percentages range between 40–60%, offtake ranges between 22–26% for animals aged between 3 to 4.5 years old and with dressed carcass weights between 225–360 kg. Most of the meat produced is exported as manufacturing beef ('t Mannetje, 1982).

By oversowing a legume into native pasture, or by establishing grass-legume pastures, substantial improvements in animal production can be obtained. Edye and Gillard (1985) reported that by oversowing *Stylosanthes hamata* cv. Verano into the native pasture along with the use of phosphate fertilizer, the carrying capacity could be increased up to ten-fold and cattle fattened in about half the time when compared with native pastures.

In an experiment at the Narayen Research Station in south-east Queensland three herds of breeding cows were used to measure beef production on unimproved native pasture at a stocking rate of 0.17 cows/ha and sown grass-legume pasture with superphosphate at stocking rates of 0.50 and 0.68 cows/ha over a period of ten years (Coates and 't Mannetje 1990). A summary of the results is given. On both treatments conception and calving rates were high, but the long term average calving percentage of 70–75% for commercial beef production in the region on native pastures is also considerably higher than the Queensland state average (c. 60 %). All production parameters on sown pasture were higher than on native pasture. The results are most strikingly presented in terms of production per ha. Sown pasture produced 4.5 times more weight of calves per ha than native pasture and they were 20 % heavier.

Africa: In Africa most livestock are kept by smallholders for numerous reasons, the most important being milk and manure production, although traction and beef production may be locally important. Also throughout Africa cattle act as capital reserves and offer security of income or survival. Animal husbandry practices differ greatly between traditionally kept herds and commercial ranching, with corresponding differences in production, which can largely be attributed to disease control and animal husbandry. Birth weight and liveweight gains in indigenous cattle are low for both production systems, but particularly for traditional systems where the sale of animals for beef is not the primary objective.

Table : *Mean conception and calving rates, average daily gain of calves, weaning weight (7 months of age), weight of calf per cow mated and weight of calf per ha on unimproved native pasture and sown grass-legume pasture in south-east Queensland.*

	Native pasture	*Grass-legume pasture*
Conception rate	84 %	90 %
calving rate	79 %	82 %
Average daily gain calves	0.78 kg	0.92 kg
Weaning weight	200 kg	241 kg
Weight of calf/cow mated	155 kg	199 kg
Weight of calf/ha	26 kg	118 kg

The Value and Marketability of Animal Products

Nomadic herdsmen and small farmers in Africa are usually subsistence producers with limited access to markets. It is unlikely that these producers will have the necessary inputs required for pasture improvement.

Large scale producers are by definition dependent on a market for their produce. Price in the local, national and international markets will be determined by supply and demand. Local demand will, in turn, depend on consumers' incomes, therefore, in most developing countries producer prices for animal products tend to be low and there is little incentive to invest in pasture improvement.

Table: *Calving rates, birth, weaning and live weights of different breeds and production systems in Africa (Brumby and Trail 1986).*

Country	Breed/system	Calving %	Weight (kg)			4 yrs
			Birth	Weaning	2 yrs	
Mali	Sudanese/Fulani					
	traditional	54	17	55	125	200
	ranching	77	21	79	220	280
Nigeria	White Fulani/					
	traditional	46	20	55	140	240
	ranching	89	24	96	245	350
Ethiopia	Boran/					
	traditional	55	20	55	150	260
	ranching	78	25	180	265	420
Botswana	Tswana/					
	traditional	46	26	120	260	300
	ranching	74	31	180	360	400

The Motivation of Farmers

An important consideration of a producer contemplating pasture improvement is the cost and the expected return on investment. Costs are high and it depends on the financial situation of the farm whether it would be attractive to apply the technology. Although, eventually, a high return may be achieved, during the first few years after pasture improvement there will be a negative cash flow (Wicksteed, 1986). Not all farmers are in a position to carry this and credit facilities are not always available. There is also a risk involved that the improvement may fail, or that beef prices will decrease.

D.B. Coates (pers. comm.) listed as the main reasons for the present interest in pasture development based on legumes in Queensland (Gramshaw and Walker 1988) as:

- better beef prices providing money for investment;
- change in management brought about by compulsory testing for brucellosis and tuberculosis;
- high cost of mustering on large properties;
- market pressures for higher quality beef and faster turnoff rate;
- more emphasis on business aspects of the grazing industry;
- simple technology of oversowing *Stylosanthes* into native pasture at relatively low cost;
- new approaches to extension; and
- "successes" are increasing and so building up confidence.

Land Tenure

Land tenure can be a severe limiting factor to pasture improvement. If the producer does not have security of tenure then it is unlikely that he will be inclined to invest money into long-term pasture improvement. Most grazing lands in Africa are open to common grazing by privately owned herds.

This system works well with low human population pressures and therefore low animal numbers in a region. However, as human and animal population increase so does the pressure on the land and, since individuals are not responsible for the control of the grazing or for the maintenance of the land, everybody tries to maximise their return from the communal resource. This leads to severe overgrazing, low productivity and danger of soil erosion. When land is owned or leased by the producer, there is the possibility of matching stocking rates with the carrying capacities in order to maintain and improve grazing and to replenish fertility. The chance of achieving a sustainable production system is less with communal than with privately owned grazing land.

The Availability of Resources and Knowledge

Before pasture improvement can be undertaken, there has to be an infrastructure for the distribution of inputs, such as, seed and fertilizers. In many regions there is no seed production or a seed trade. Pasture improvement is further constrained by the lack of finance and of knowledge by the farmers and the extension service. It is therefore necessary to promote the extension of pasture improvement practices, to encourage the development of seed production and to educate bankers on the possibilities of pasture improvement.

Opportunities for Forage Production Improvement in Different Livestock Systems

Forage Production Systems

Based on a classification of Perkins *et al.* (1986) for Indonesia and modified by 't Mannetje and Jones (1989) for south-east Asia; a number of distinct forage production systems can be distinguished on a global scale.

- Extensive permanent grasslands are unimproved native pastures (rangelands), which receive no inputs such as irrigation or fertilizer. On privately owned or controlled land, management consists of controlled grazing, burning and the control of trees and shrubs. Whereas communally grazing, which includes the savannas in Africa as well as road/canal sides and open forests, there are no forms of control and there is frequently overgrazing.

- Semi-intensive forage production systems have inputs, such as, fertilizer and weeding also irrigation (primarily applied to the main cash crop) which also benefits associated forage production.
- Intensive forage production systems have inputs applied for the sole purpose of forage production.

The ability to improve forage production is a function of climate, soils and the production system. Water and nutrients are physical prerequisites. Under natural conditions improved forage production is possible where rainfall is in excess of 650 mm/yr with a dry period not exceeding six months. In drier regions irrigation can be used. The existing production system, particularly the type of land tenure, also determines whether forage production can be improved and the type of inputs that can be realised.

Livestock Production Systems.

Livestock production takes place in three main systems:

- Pastoralism (nomadism, ranching, dairying)
- Livestock-crop systems
- Crop-livestock systems

Pastoralism: The main feature of this system is that the producers are entirely dependent on livestock for all their needs, there is no supplementary food crop production. There are three contrasting systems within pastoralism, *viz.* nomadism, ranching and commercial dairying.

Nomadism: is the way of life of indigenous peoples in arid and semi-arid regions who have no permanent place of settlement and move with all their livestock and possessions in search of water and forage. Regions in which nomadism is practised are characterised by an environment where the primary productivity which is so low that people cannot avail themselves of their feed requirements within a day's reach of a permanent settlement. The animals provide blood, milk, meat and income from the sale of surplus stock. The main source of animal feed is derived from extensive, unimproved, communally grazed grasslands. Post-harvest nomads are allowed to graze their animals on stubbles to consume crop residues and weeds and to deposit manure. However, nowadays, most crop farmers own livestock themselves and nomads are becoming less welcome. There are no real means to improve the feed base since the land is in communal resource.

Ranching: In Australia, Africa and South America ranching is undertaken in the sub-humid, semi-arid and arid regions to produce beef and wool. Properties are often large (1,000 ha to thousands of km^2), particularly in semi-arid and arid regions and may carry from 500 to several (20–50) thousand cattle or sheep and are usually either on long term leases or freehold. Production is extensive, with low inputs and corresponding low productivity both per animal and per unit area. The main feed supply are extensive unimproved native grasslands. However, ranching has the potential for sustained use of such grazing lands plus the opportunity to improve part of it by either establishing grass-legume pastures, by oversowing with a legume or by establishing a protein or fodder bank.

The area for improved forage production will usually be small compared to the total area of the property and a judicious management system is necessary to optimise animal production and to increase the sustainability of the production system (Rickert and Winter, 1980; Coates and 't Mannetje, 1990). The extensive unimproved grassland can be used as the basic forage resource, with supplementation from small improved areas of grass-legume mixtures, protein banks or fodder crops for special classes of livestock (breeders, steers for fattening) at times of nutritional stress.

Larger areas of legumeoversown grasslands can be used to either increase the overall carrying capacity or to reduce the area of land required for the existing herd (Edye and Gillard, 1985).

The reasons for improving forage production are to supplement the basic forage resource with more and/or better quality feed for various purposes, which include:

- to reduce mortalities which occur during dry times and where extra feed from improved pastures, protein banks or fodder crops can be used either for the whole herd to provide (near) maintenance rations or for a nucleus of animals to ensure continuation of the herd;
- to increase the conception and calving rates where an increasing plane of nutrition leading up to conception can increase fertility and reduce calving intervals;
- to keep the herd on a smaller area of land during the whole year, or part of the year (by oversowing part of the area) in order to reduce mustering costs; and
- to fatten younger stock on improved pasture or on feed from fodder crops or protein banks.

Table: *Forage production systems used in different livestock production systems.*

	LIVESTOCK SYSTEM			NRDLC	
	O	A	A	I	R
	M	N	I	V	O
	A	C	R	.	P
	D	H	Y	C	-
	I	I	I	R	L
	S	N	N	O	I
FORAGE PRODUCTION SYSTEM	M	G	G	P	V
EXTENSIVE PERMANENT GRASSLAND					
privately used	-	+	+	-	-
communally used	+	-	-	+	+
SEMI-INTENSIVE PERMANENT FORAGES					
understorey tree crops	-	-	+	-	+
forage from shade trees	-	-	+	-	+
forage in alley cropping	-	-	+	-	+
edges of crop fields	+	-	+	+	+
SEMI-INTENSIVE ANNUAL FORAGES					
after harvest crop	-	-	-	+	+
crop residues	+	-	+	+	+
INTENSIVE PERMANENT FORAGES					
improved grasslands	-	+	+	-	-
protein banks	-	+	+	+	+
INTENSIVE SHORT-TERM FORAGES					
fodder crops	-	+	+	+	+

Commercial milk production: is practised in regions of better rainfall and with access to markets, mostly near to urban centres. In humid regions dairy farms use improved pastures consisting of tropical species whilst in subtropical regions winter pastures consisting of irrigated temperate species may be appropriate (Stillman *et al.*, 1984). On tropical pastures milk production never exceeds 15 kg milk/cow/day without the use of concentrates and usually it is less than 10 kg; whilst on unimproved grasslands production is very low (3–5 kg milk/cow/day) (Stobbs and Thompson, 1975).

Livestock-crop systems: In regions with better rainfall, instead of nomadism people obtain their livelihood from a combination of livestock rearing and cropping. In these transhumance or seminomadic systems people have permanent settlements with some food cropping on better soils and a few animals for susistence, but with the main herd being moved to distant grazing lands in search of water and forage and returning to the settlements in the wet season. The main feed is obtained from extensive communally grazed grasslands yet there are opportunities to establish protein banks and fodder crops on privately owned crop land which also supply crop residues.

Crop-livestock Systems: These occur in regions with relatively high rainfall or a reliable wet season. The main activity is crop production although there may also be a high domestic livestock population. In the humid tropics rice and, in drier regions, other cereals such as millet, sorghum or maize are the main food crops. Because of high human population densities there is often a land shortage and farms are usually small therefore crop production is intensive. Animals (cattle and buffaloes) are needed for draught power, extra income, food and dung which may be used as either a fertiliser or fuel.

The forage for the animals is obtained from waste land, road sides or canal banks. In some countries (e.g. Malaysia) there are publicly owned grazing reserves or areas of abandoned crop land, often covered in weed grasses such as *Imperata cylindrica*, are used as grazing lands (as in The Philippines and Indonesia). In west Africa grazing land is often fallow crop land.

Tree crop plantations offer great opportunities for the production of forage, particularly under mature coconuts and newly planted oilpalm before the canopy is completely closed (Reynolds, 1988). In Indonesia coffee plants are shaded by *Leucaena leucocephala* and pruned branches provide fuelwood and leaves for animal feed. Similarly, alley farming also offers opportunities for livestock feed production (Sumberg and Atta-Krah, 1988). Crop residues are important as forage and crop by-products are available as concentrate feed.

There are few other opportunities for improving forage production because of the lack of land, however, leguminous trees and shrubs can be used as living fences or as single trees in back yards. Depending on the need for land for food or cash crops and the price of livestock products, it may be profitable to grow fodder crops.

Table 4 shows which forage production systems can be used in the different livestock systems. This indicates clearly that commercial dairying and crop-livestock systems have the greatest range of forage production systems at their disposal and that nomadism has the least options.

In each production system there is a basic resource which provides the bulk of the forage. In nomadic and ranching systems, as well as in livestock-crop systems, the unimproved extensive grasslands, which are too poor for cropping, provide the basic forage although yields are low. There is no point in fertilising native vegetation since the response will not repay the cost of the fertiliser (Henzell and 't Mannetje, 1980). On the other hand, there is little export of nutrients from extensively grazed pastures and management must aim at sustainability through controlled grazing, burning and control of woody weeds.

Extensive grasslands are, however, the last major land resource in many regions of the world and with increasing population pressure these grasslands are being converted into marginal crop lands - such areas are liable to nutrient exhaustion and erosion. Controlled land use will remain an impossibility whilst there are communal rights of access because it is not in any individual's interest to reduce his use on the land if others continue or even increased their usage. The only solutions are controlled use and management of the grazing lands either instigated by the community using it or by government and population growth control.

Low cost forage improvement technology based on legumes with minimal fertiliser input is available. However, in developing countries serious constraints include communal land use, the low value of animal products and the lack of knowledge and resources, including finance. Integrated systems of crop and forage production and integrated use of forage production systems need to be further developed. The extensive unimproved grasslands (rangelands) are a precious resource which is under pressure of overuse and conversion to crop land.

Table: *Selected grasses and legumes for pasture improvement in four climatic regions as defined by Troll (1966) ('t Mannetje and Jones 1989)*

Grasses	***Climatic Zones***			
	V_1 & V_2	V_3	IV_7	IV_4
Andropogon gayanus	x	x	-	-
Brachiaria brizantha	x	x	-	-
Brachiaria decumbens	x	x	-	-
Brachiaria dictyoneura	x	x	-	-
Brachiaria humidicola	x	x	-	-
Brachiaria mutica	x	x	-	-
Brachiaria ruziziensis	x	x	-	-
Cenchrus ciliaris	-	x	-	x
Chloris gayana	-	x	x	x
Cynodon dactylon	x	x	x	-
Cynodon nlemfuensis	x	x	-	-
Digitaria decumbens	x	x	x	-
Digitaria setivalva	x	x	-	-
Panicum maximum	x	x	-	-
Panicum maximum var. trichoglume	-	-	x	x

Contd...

Grasses	***Climatic Zones***			
	V_1 & V_2	V_3	IV_7	IV_4
Paspalum dilatatum	-	-	x	-
Paspalum plicatulum	x	x	x	-
Pennisetum clandestinum	-	-	x	-
Pennisetum purpureum	x	x	-	-
Setaria sphacelata	x	x	x	-
Sorghum almum	-	-	-	x
Sorghum sudanense	x	x	x	x
Tripsacum laxum	x	-	-	-
Urochloa mosambicensis	-	x	-	-
Zea mays	x	x	x	-
Shrub Legumes				
Albizia lebbeck	-	x	-	x
Calliandra calothyrsus	x	-	-	-
Codariocalyx gyroides	-	x	x	-
Flemingia macrophylla	x	-	-	-
Gliricidia sepium	x	-	-	-
Leucaena leucocephala	x	x	x	x
Sesbania grandiflora	x	-	-	-
Sesbania sesban	x	-	-	
Herbbaceous Legumes				
Aeschynomene americana	x	-	x	-
Aeschynomene falcata	-	-	x	x
Alysicarpus vaginalis	x	x	-	-
Arachis pintoi	x	x	-	-
Calopogonium mucunoides	x	x	-	-
Cassia rotundifolia	-	x	-	x
Centrosema acutifolia	x	x	-	-
Centrosema macrocarpum	x	x	-	-
Centrosema pascuorum	-	x	-	-
Centrosema pubescens	x	x	-	-
Clitoria ternatea	-	x	-	-
Desmodium heterocarpum	x	x	x	-
Desmodium heterophyllum	x	x	-	-

Contd...

Grasses	***Climatic Zones***			
	V_1 & V_2	V_3	IV_7	IV_4
Desmodium intortum	-	-	x	-
Desmodium ovalifolium	x	x	x	-
Desmodium triflorum	x	-	-	-
Desmodium uncinatum	-	-	x	-
Lablab purpureus	x	x	x	x
Lotononis bainesii	-	-	x	-
Macroptilium atropurpureum	-	x	x	x
Macroptilium lathyroides	-	x	x	x
Macrotylome axillare	-	x	x	x
Medicago sativa	-	-	x	x
Mimosa pudica	x	-	-	-
Neonotonia wightii	-	-	x	-
Stylosanthes capitata	x	x	-	-
Stylosanthes hamata cv. Verano	-	x	-	x
Stylosanthes humulis	-	x	-	x
Stylosanthes guianensis	x	x	x	x
Stylosanthes macrocephala	x	-	-	-
Stylosanthes scabra	-	x	-	x
Trifolium repens	-	-	x	-
Trifolium semipilosum	-	-	x	-
Vigna parkeri	-	-	x	-
Vigna unguiculata	x	x	-	x

Practical Technologies for Mixed Small Farm Systems in Developing Countries

In animal production systems, sustainable agricultural practices are those that promote systems in which there is efficient use of the natural resources based on an understanding of the interrelationships within prevailing agro-ecosystems. These need to be identified with agricultural and rural development processes that have an integrated approach to natural resource management in which potentially important technologies are technically appropriate, economically viable and socially acceptable to target beneficiaries, the small farmers. In small farm systems, the objective of achieving sustainable animal production involving crops and animals assumes two major considerations. Firstly, it is essential that the available animal genetic resources within the mixed small farm systems will be fully exploited in a manner that is consistent with their biological attributes and potential productivity. Secondly, the choice of practical technologies should be realistic of the needs of such systems in order to promote sustainable productivity, as well as provide economic stability.

The prevailing circumstances suggest, however, that these factors are far from being in place and is reflected by the a) inefficient use of ruminants (buffaloes, cattle, goats and sheep), b) few examples of systems that are demonstrably sustainable, and c) concurrent low animal productivity. The problems are complex especially in small farms. These farms are limited by both access and availability to adequate resources.

The situation is further exacerbated by the growth of both human and animal populations, rural poverty and environmental degradation; all of which emphasise that the choice of practical technologies are especially important and can considerably influence the sustainability of the system. Another dimension concerning sustainability is the livelihood of the farmers themselves. In extreme situations where land sizes are particularly small or where landless labourers or tenants are involved, ownership of animals provides the only means of sustaining their livelihood. Nomads, pastoralists and transhumant peasants fall into this category. In all regions, but notably in north Africa, Middle East and north Pakistan, India and China, to whom ownership of animals is the key to livelihood and survival. In these situation, sustainable practical technologies for animals assumes even greater significance in socioeconomic terms.

It is appropriate therefore to examine the role of animal production in small farm systems and, in particular, to identify practical technologies that can significantly contribute to more sustainable systems and environmental integrity. The choice of technologies needs to take cognizance of the importance of crop-animal interactions where both components are complimentary and beneficial in terms of productivity from the land.

The purpose of this paper is to focus on those technologies which have been used in various developing countries and which show potential value. Specific case studies are cited in Asia, Africa and Latin America which highlight the relevance of these technologies in small farm systems involving crops and animals.

Size of Holding

Farm size is an important consideration and actual size varies between and with regions and countries. In Africa, North and Central America and Asia, the highest percentage of holdings are less than one hectare in extent.

In Asia and Asia the majority of holdings are less than 5 ha. The smallest farm size occurs in Bangladesh, while households cultivating paddy in Sri Lanka have average holdings of 0.3 ha of land. Associated with the size of small farms, is the marked imbalance between the total ruminant livestock units and the available permanent pastures between regions.

In Asia including centrally planned economies, the situation is especially critical where livestock densities are the highest in the world (FAO, 1986).

These small farms constitute the backbone of traditional agriculture throughout the developing countries (Devendra, 1983). Small size presents a serious constraint, partly because of lack of access to resources, but more particularly, because of the difficulties of promoting the application of practical technologies capable of a discernable impact.

Thus, the small size of the farms determine, to a large extent, by the scale and extent to which new technologies can be adopted, and this aspect needs to be considered in the formulation and application of suitable technologies.

Practical Technologies

Notwithstanding holding size as a constraint to production and sustainability, there exist a number of practical technologies that have been tried with varying degrees of success. Really good examples of proven and demonstrable technology are, however, few.

This is partly because they have been inadequately tested on-farm with farmers, but more particularly, because chosen technologies have not been considered in a holistic and systems context.

Nevertheless, several potentially valuable technologies are apparent and these will be discussed in the following sections and can grouped into two broad production systems:

- situations involving arable cropping systems, and
- situations involving tree cropping systems.

***Table:** Distribution of Size of Holdings by Region in the Developing Countries (Fao, 1971)*

Region (number of reporting countries is given in parenthesis)	Total number of holdings (10^3)	% Distribution									
		Holdings without land*	Under 1 ha	1–2 ha	2–5 ha	5–10 ha	10–20 ha	20–50 ha	50–100 ha	100 – 200 ha	200 + ha
Africa (20)	13,110.9	1.6	39.8	27.8	22.5	5.5	1.7	1.0	0.1	0.0	0.0
N & C America (14)	6,008.0	2.6	18.6	8.8	9.9	6.8	7.5	14.0	13.1	9.7	0.9
South America (7)	8,383.6	0.2	15.5	12.5	20.5	13.7	12.7	12.6	5.3	3.2	3.7
Asia (16)	103,619.8	0.2	52.7	19.1	17.7	6.6	2.7	0.8	0.1	0.01	0.0

* Establishments with no agricultural land which raise livestock and livestock products

***Table:** Variation in the size of small farms in some countries in South East Asia (FAO/UNDP, 1976)*

Country		***Definition***
Bangladesh	a)	Subsistence farmers-cum-sharecroppers < 0.4 ha.
	b)	Viable and potentially viable owners, 0.4 to 0.8 ha.
India	a)	Small farmers, 2 to 4 ha of dryland (1 ha wet = 0.8 ha of dryland) and whose annual income <Rs 2400⁺.
	b)	Marginal farmers, 0.8 to 2 ha of dryland and annual income <Rs 1200.
	c)	Agricultural labourers, < 0.8 ha of dryland and annual income <Rs 1200.
Indonesia	a)	Average size was 1.7 to 2.0 ha.
Korea	a)	Less than 1 ha and income below 500,000 Won per annuM⁺⁺.
Nepal	a)	Terai, 4 bighas (2.5 ha).
	b)	Hills, 1.75 bighas (1.0 ha).
Philippines	a)	Average size ranges between 0.8 to 7 ha.
Sri Lanka	a)	Agricultural households, 1.2 ha of land.
	b)	Paddy cultivating households, 0.3 ha of land.
Thailand	a)	Non-canal-irrigated areas: 15 rai*. canal-irrigated areas: 10 rai and net cash income of 1000 Baht**.

[+] US$ = 8.50 Rs (approximately)

[++] US$ = 285 Won (approximately)

* 1 rai = 0.16 ha

** US$ = 26 Baht (approximately)

Situations Involving Arable Cropping Systems

The Three Strata Forage System

In dryland farming areas, a major constraint to higher productivity from ruminants is the unavailability of good quality feeds especially during the dry season and periods of drought. The development of feeding systems that can increase the supply of good quality forages are, therefore, especially important to improve the prevailing low level of animal performance. This is exemplified by the situation in Bali which has approximately three million people and three rainfall zones. Twenty five per cent of the land area is semi-arid with a rainfall of 900–1500 mm/yr. Farmers constitute about 70% of the total population and most of them practice mixed crop-animal farming.

Among the ruminants, Bali cattle are particularly important and, in the dry parts of the island, income from livestock accounts for between 29–43% of total farm income. Farmers generally own 2–3 herd of cattle which are used for draught and beef production. The feed resources for ruminants in Bali come mainly from native grasses, tree leaves and cereal straws. The dry matter productivity from these resources is generally very low (approximately 2–2.5 tonnes/ha/yr). Dry matter yields can be increased through the introduction of improved grasses as well as forage legumes, eg. leucaena (*Leucaena leucocephala*) and *Gliricidia sepium*.

Bali cattle generally feed on wayside grasses and crop residues. Live weight gains vary between 100– 200 g/day and marketable weights are only reached in about four to five years.

However, with improved feeding and concentrate supplementation, it has been shown that daily weight gains of Bali cattle could be increased to between 400–600 g/day and that the fattening period could be reduced to less than two years. These circumstances led to the development and successful demonstration of the Three Strata Forage System, a project supported by the International Development Research Centre (IDRC) of Canada. The system involves a first stratum of grasses and ground legumes; a second stratum of shrub legumes; and a third stratum of fodder trees. The project ran for five and a half years and several relevant results were found comparing two types of systems:

The Three Strata Forage Systems (TSFS) and the non-TSFS (NTSFS) at two stocking rates (2 and 4 cattle/ha). The results of the project have been published (Nitis *et al.*, 1990; Lana *et al.*, 1990; Arga, 1990; and Nuraini, 1990).

Some of the results, and the following summarises the main highlights:

- The allocation of 0.09 ha of land as a forage boundary within the TSFS produced 98% more forage than the 0.25 ha of natural pasture. With the additional 2000 shrubs and 42 trees, the total wet and dry seasons forage production in the TSFS was 91% more than that of NTSFS.
- *Stylosanthes, Centrosema, Acacia, Gliricidia and Leucaena* which were grown, provided increased dietary nutrients in the TSFS. Consequently, cattle raised in TSFS gained 19% more live weight and reached market weight 13% faster.
- The availability of increased forages also enabled higher stocking rates and live weights to be achieved: 3.2 animal units (375 kg)/ha/year in the TSFS compared to 2.1 animal units (122 kg)/ha/year.
- Cattle in the TSFS were less infested by endoparasites. This was presumably due to less contact with the traditional cattle, since the TSFS cattle were always kept in confinement.
- The introduction of forage legumes into the TSFS reduced soil erosion by as much as 57% in the TSFS compared to NTSFS. Additionally, the fertility of the soil in the TSFS was also considerably improved.
- With the presence of 2000 shrubs and 42 trees lopped twice a year, firewood production from 0.25 ha of TSFS was 1.5 tonnes per year and TSFS supplied 64% of the farmer's requirement.
- Farmers in the TSFS spent less time managing their cattle and, more importantly, these same farmers benefitted by an increased 31% income compared to NTSFS farmers.

Further research and development aimed at creating a more sustainable TSFS will involve the introduction of goats. The justification for including goats is a) the official promotion of goats in dryland farming areas, and b) the potential for generating additional income.

Additionally, goats will provide greater flexibility of resource use by farmers. On a live weight basis a 375 kg Bali bull is equivalent to 6 30 kg goats. When feed is limited during the dry season, it is easier to reduce the number of goats rather than cattle.

Theoretical calculations suggest that with shrubs and tree fodders, increased carrying capacity is feasible and detailed studies are continuing on the utilisation of *Gliricidia* as a forage.

Table: *Comparative Productivity of TSTS and NTSFS Plots (kg dry weight/plot per year) (Nitis* et al.*, 1990)*

Parameter	***TSFS***[+]	***NTSFS***[++]
Food	853	1268
Straw	750	1218
First stratum	455	-
Second stratum	310	-
Third stratum	15	-
Shrubs	-	132
Trees	-	2
Improved grasses	-	10
Native grasses	-	242
Firewood	1049	475
Cattle live weight gain (kg/3 years)	186	166
Carrying capacity (cattle/ha)	4	2
Maximum live weight (kg/head)	300	200
Soil erosion (mm/2 years)	11	20

[+] Three strata forage system

[++] Non-three strata forage system

Food-Feed Intercropping

The concept of food-feed intercropping in both lowland and upland small farm systems is relatively new. The two principal advantages are: a) that the system aims to provide sustainability through involving the complimentary role of crops and animals; and b) the use of appropriate forage crops provides fodders and crop residues which are valuable both ruminants and non-ruminants. Already several attempts to develop food-feed systems have been undertaken notably in the Philippines and Thailand. Rice is the principal cereal crop, but other crops which have been used with the aim of increasing the production of animal feeds include: cowpea, maize, groundnut, pigeon pea, sorghum and sweet potato. The criteria for the choice of the inter-crop include *inter alia*: the type of animals reared, potential forage or crop residue yield, promotion of soil fertility, drought tolerance and extent of the dry season, shade tolerance in upland areas, ease of eradication and resource requirements. The strategy is to integrate within the rice cropping pattern (intercropping and relay cropping) other feed producing crops and forage crops without reducing the area of the land used. Exemplifies this situation in the Philippines. In the past, the general tendency has been to grow only one crop of rice in both the rainfed lowland and upland areas resulting in a low cropping intensity. Both areas are important for livestock production. Furthermore, the rainfed lowland rice areas occupy about 67% of the total land area in Asia and where the bulk of swamp buffaloes, cattle and sheep are found.

The drier upland areas generally favour the presence of small ruminants, mainly goats, but also some large ruminants. Rice and wheat straws are the main residues from cereal cultivation and these supply the basic diet for ruminants. They provide bulk and energy for maintenance but not production. The intake of these roughages is limited by their low crude protein content (4–6%) and low digestibility, which necessitates supplementation to meet production requirements. Additionally, the reduced feed availability during the dry season and associated weight loss and poor performance, necessitates the application of strategies to ameliorate the situation and increase available dietary nutrients. In the arable areas, cereal straws (mainly rice and wheat) and other crop residues are the most available and cheapest feeds for ruminants. The greatest challenge rests with the development of effective and economic feeding systems which utilise the high lignocellulosic content of straws.

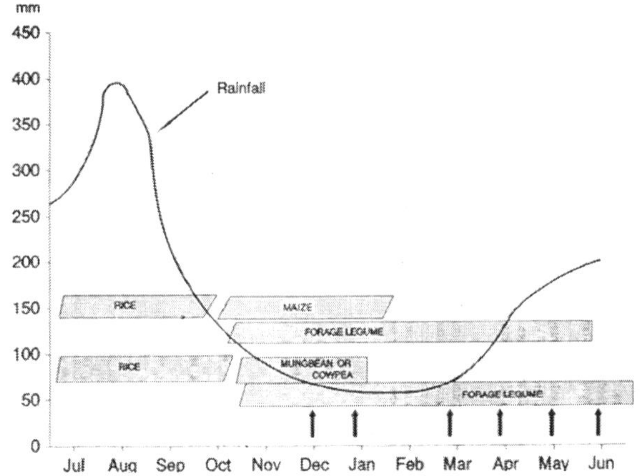

Figure: *Cropping patterns involving rice and food crop - forage intercropping*

The value of integrating two forages lablab (*Lablab purpureus*) and *Clitoria ternatea*.

The latter is grown by farmers in northern Philippines after rainfed lowland rice as monocrop for green pods and forage. No differences were found in mungbean grain and residue yield. The combination of mungbean plus lablab (cv Rongai) gave the highest forage yield of 7.9 tonnes/ha. Similarly, the results of an experiment on maize-forage intercropping under rainfed upland conditions (Tengco and Carangal, 1987) involving varying plant densities. Four forage legumes were used: *Stylosanthes guianensis* CIAT 136, *Macroptilium atropurpureum* cv Siratro, *L. purpureus* cv Highworth and *Desmanthus virgaties.* Of these, Siratro gave the highest forage yield. The results demonstrate the value of intercropping with suitable forages to produce valuable feeds for animals.

Table: *Grain, Residues and Forage Yields of Mungbean and Forage Legumes as Monocrop and Intercrop Combinations in the Philippines (tonnes/ha) (carangal* et al., *1988)*

Crop Combination	Grain	Residues (DM)	Forage Yield (DM)					Total (DM) forage & residue
			Initial cut	Regrowth			Total	
				1	2	3		
Mungbean + lablab (cv. Hatiya)	1.02	0.91	1.69	1.98	1.11	0.91	5.69	6.60
Mungbean + lablab (cv. Rongai)	1.00	0.94	1.99	2.43	1.83	1.68	7.93	8.87
Mungbean + clitoria	1.28	1.14	-	1.53	1.16	0.93	3.62	4.76
Mungbean	1.26	1.19	-	-	-	-	-	1.19
Lablab (cv. Hatiya)	-	-	2.70	1.73	1.16	1.40	6.99	6.99
Clitoria	-	-	2.43	2.70	2.06	1.73	8.92	8.92
CV (%)+	18	13	20	32	20	32	10	
LSD++	NS	NS	0.89	1.17	0.49	0.71	1.79	

+ Coefficient of variation

++ Least significant difference (P < 0.05)

Table: *Yields of Maize and Forage Legume Intercropping at Three Plant Densities during the Dry Season in the Uplands, Philippines (tonnes dm/ha) Tengco and Carangal, 1987)*

Intercropping	Main crop		Intercrop
	Grain	Fodder	Forage
Maize 26,666 pl/ha monocrop	2.95	3.72	-
+ Stylo CIAT 136	3.14	3.91	1.84
+ Siratro	2.57	2.57	2.95
+ *L. purpureus*	3.19	3.02	1.54
+ *D. virgatus*	2.69	3.21	0.64
Maize 35,555 pl/ha monocrop	2.94	3.62	-
+ Stylo CIAT 136	2.27	3.36	1.14
+ Siratro	2.41	3.77	2.88
+ *L. purpureus*	2.19	3.28	1.14
+ *D. virgatus*	2.22	3.15	0.62
Maize 53,333 pl/ha monocrop	4.40	4.94	-
+ Stylo CIAT 136	4.23	5.09	0.60
+ Sirato	3.40	4.49	2.72
+ *L. purpureus*	3.70	3.89	1.69
+ *D. virgatus*	3.75	5.17	0.75
C.V. (%)±	28.4	23.4	47.2
F-test	**	**	**

+ Coefficient of variation

++ Level of significance

Relay Cropping

Relay cropping is an important means to increase the supply of feed for farm animals. In relay cropping, a second crop is planted into the first before harvest eg. the introduction of legumes (groundnut or pigeon pea) into a main crop of rice or wheat. The strategy can extend the supply of feed, possibly, throughout the year.

Integrated Pig-Ducks-Fish-Vegetable Systems

A sustainable system which is widely practised in South East Asia and China involves the integration of pig production with fish farming, duck keeping and vegetable production, or a combination of these (Devendra and Fuller, 1979). The inter-relationships between the components. The system is based on the use of ponds which not only meet the needs of pigs, but also enables fish and ducks to be kept. Water is also useful for vegetable production.

Such systems are especially suited for small farms where only a few pigs are reared together with either ducks or fish and intensive vegetable production. Edwards (1983) estimated that 26.7 laying ducks or 8.2 pigs or 0.8 dairy cow or 1.7 buffaloes are required to produce a mean yield of 174.7 kg of fish/year in a 200 m^2pond, based on equivalent nitrogen manure inputs.

Situation Involving Tree Cropping Systems

Integrated Animal and Tree Cropping Systems

Integrating ruminants with tree cropping systems, such as, coconuts, oil palm and rubber are important production systems which have not been adequately exploited. Livestock/tree cropping systems are common in the humid and subhumid regions where intensive tree-crop production is practised. Although such systems are not new greater attention is needed to ensure more complete utilisation of the land. The advantages of the system are:

- increased fertility through recycling of nutrients (dung and urine)
- weed and shrub control
- reduced use of herbicides
- reduced fertiliser wastage
- provision of shade
- easier management of the plantation, and
- distinct possibilities of increased crop yields and animal products from the land.

There is an estimated area of 20.3×10^6 ha under tree crops in South and Southeast Asia which reflects the potential for this kind of activity (FAO, 1986). Many of the Pacific island territories, notably Papua New Guinea, New Hebrides, Fiji, the Solomon Islands and Western Samoa have large areas under coconuts and a potential for integrating ruminant production. Malaysia provides a specific example of the economic benefits of integrating ruminants with oil palm where an estate has allocated a portion of the plantation to the workers for grazing their animals. For the first two years (1980 and 1981) only cattle were owned and grazed, however, in 1982 and 1983 goats were also introduced. This was done because of their economic importance and capacity to supply both meat and milk in the estate.

A comparison between the grazed and non-grazed area involving both young and mature trees is valid in that it involved both the same area of 71–135 ha and, more particularly, the fact that both areas were on the same soil type. The effect grazing cattle and goats was an increase 2.15–5.16 tonnes fresh fruit bunches per hectare per year over four years. When translated into the total hectare available for grazing and sale value per tonne of fresh fruit yield, the economic advantage is substantial. The result in economic terms is similar to the findings in West Java of integrating sheep and goats under rubber. The presence of legume is of definite advantage and it has been calculated that the amount of nitrogen utilised by the animal and excreted in the faeces and urine increases with the presence of the legume cover.

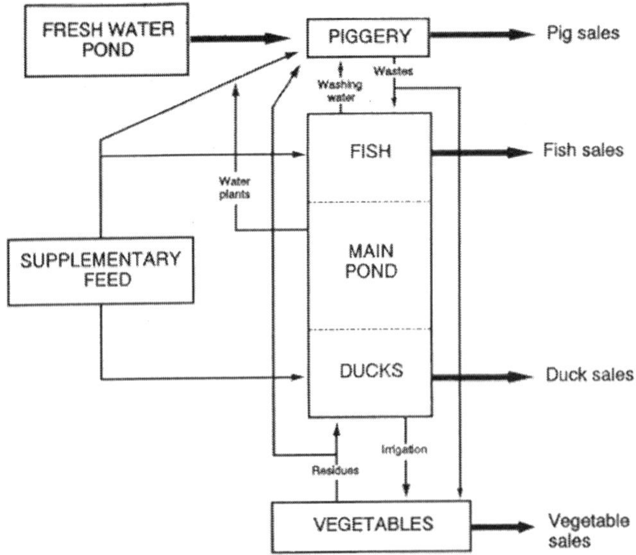

Figure: *Integrated pig - fish - duck - vegetable system*

***Table:** Effect of Mixed Cattle and Goat Grazing on the Yield of Fresh Fruit Bunches in Oilpalm Cultivation in Malaysia (Devendra, 1986)*

Year	Yield of fresh fruit bunches (tonnes/ha)		
	Annual grazed area	Annual non-grazed area	Difference
1980	30.55 (cattle)	25.61	4.94
1981	17.69 (cattle)	15.87	1.82
1982	25.12 (cattle and goats)	22.97	2.15
1983	23.45 (cattle and goats)	18.29	5.16
Mean	24.20	20.69	3.51

More recently, a study has reported the integration of goats with pine (*Pinus insularis*) in the Philippines (Penafiel and Veracion, 1987). A stocking rate of four goats/ha did not cause significant soil loss nor disturbance of soil bulk density, compaction and infiltration rates. Thinning of the pine stands increased tree growth as well as forage production. The daily weight gains were impressive and averaged 99 to 129 g/day/goat. The integration demonstrated a ecological and economically viable and sustainable farming practice for the forest dwellers.

Alley Cropping

Alley cropping provides another opportunity to integrate crops and animals as well as enhancing sustainability. The development of the system has the following advantages:

- maintenance and promotion of soil fertility
- increase the supply of animal feeds
- supply of fuelwood, and
- contribute to the development of all year round feeding systems.

The results of a study in Nigeria involving leucaena (*L. leucocephala*), maize and grazing with goats and sheep. The effects of both alley farming and alley farming after fallow on the yield of maize grains were 30–70% higher than that of conventionally cropped plots. The results did not separate out however, the specific effects of the presence of animals and fallow.

***Table:** Yield of Maize Grain in Conventional Farming, Alley Farming and Alley Farming after a Two-Year Grazed Fallow in Nigeria (tonnes/ha) (Reynolds and Attah-Krah, 1989)*

Farming system	*First season*	*Second season*	*Total*
Conventional (no trees)	2.13	0.93	3.06
Continuous alley farming[+]	2.41	1.70	4.11
Alley farming after fallow	3.30	2.04	5.34

[+] Involving *L. leucocephala* and grazing by goats and sheep

Supplementation

In many developing countries, regular feed shortages and droughts are common. In such conditions, subsistence feeding mainly on cereal straws results in reduced live weight and perpetual animal low productivity. Inadequate nutrition is also associated with delayed age at first parturition, increased parturition intervals, prolonged non-productive life and high mortality.

Strategic supplementation of energy, proteins and minerals offer an important means to ensure that animal performance is not reduced, especially during critical periods of feed shortages. Several alternative strategies have recently been pursued, with the objective of better utilising low-quality forages for production (meat, milk or draught). Foremost in these initiatives are a variety of chemical pretreatments (Jackson, 1977; Ibrahim, 1983; Sundstol, 1984; Doyle and Pearce, 1985). Among these, urea treatment is currently the most widely used to improve nutritive value of straws. The result has been a shift from a subsistence to a maintenance levels of nutrition, achieved by increased feed intake and/or digestibility (Ibrahim *et al.*, 1984), and occasionally growth (Perdok *et al.*, 1984; Verma, 1983) and milk production (Davis, 1983) responses.

The use of forage supplements has been secondary to chemical pre-treatments and has been underestimated and not given adequate research and development attention. For a variety of reasons, this approach has enormous potential for ruminants, especially in situations where animals are abundant and varied. It is appropriate, therefore, to review current understanding of the use of forage supplements and the benefits of this strategy. There are many advantages concerning the use of these forages (Devendra, 1988) especially in mixed farms in the humid tropics:

- availability in the farms
- accessibility
- provision of variety in the diet
- source of dietary nitrogen, energy, minerals and vitamins
- laxative influence on the alimentary system
- reduction in the requirements for purchased concentrates, and
- reduced cost of feeding.

There are several good examples of such forages, especially legumes, that can be increasingly used in small farm systems. Leucaena, for example, provides a valuable source of protein, energy and sulphur for the rumen bacteria, as well as being used for fencing, fuel and mulch. Further important examples include: *Acacia* spp., *Ficus* spp., cassava (*Manihot esculenta* Crantz), *erythryna* (Erythryna spp.), gliricidia (*Gliricidia spp.*), leucaena (*L. leucocephala*), pigeon pea (*Cajanus cajan*) and Sesbania (*Sesbania spp.*).

The potential value of these forages has recently been reviewed by Devendra (1990) and a number of observations are worthy of mention:

- With all ruminants the use of various forage supplements consistently increased live weight gain or milk production,
- in many instances the beneficial response (meat and milk) was associated with a reduced production costs,
- of the forage supplements used, legumes have been particularly advantageous, and
- in the majority of situations stall-feeding or cut-and -carry systems are most common than grazing situation,

With regard to large ruminants, the research on the value of forage supplements for draught purposes is sparse. In Thailand, the effects of breed (Murrah X Swamp and Swamp) and feed supplement (with or without) on draught (work or no work) was studied over six months. The concentrate supplement consisted of cassava chips and dried leucaena leaf in a 3: 1 ratio and was applied at 1.5 kg/head per day. The animals without supplements were feed on silage. Over the first four months before the draught capacity was assessed supplementation, as one would expect, significantly stimulated growth rate in both breeds. Concerning working ability, Thai swamp buffaloes ploughed more than the crossbred Murrah, and supplementation increased the area ploughed. However, there was no significant breed effect. Neither were differences were found in the speed of ploughing between breeds or supplementation, although, there was a tendency towards faster ploughing for both breeds when feed supplements were used (Konanta *et al.*, 1986).

In terms of practical application, a recent review of the benefits of including high protein forages supplements to the diet suggest the following optimum dietary levels (Devendra, 1988):

-	optimum dietary level on DM basis:	30–50%
-	as % of live weight:	0.9–1.5%

Table: *Wong* et al, *1987 clearly shows that Leucaena supplementation increases milk production and reduces feed costs.*

All Year Round Feeding Systems

The strategy to increase feeds within small farm systems should have the final objective of developing sustainable all year round feeding systems appropriate to the prevailing situations. In this quest, maximising feed production is essential and the following approaches are feasible:

- intercropping with cereal crops
- relay cropping
- food-feed cropping systems
- intensive use of available crop residues
- forage production on rice bunds
- alley cropping, and
- three strata forage system in dryland areas.

The key elements in the use of these approaches are the quantification of the feeds produced throughout the year and efficiency with which they are utilised by the available animals. The former enables the determination of feed balance sheets, the extent to which animals can be supported and identify the critical periods during which feed deficits need to be corrected.

The scope for increasing the efficiency of feed utilisation in innovative systems is enormous. More innovative feeding practices are necessary that can sustain all year round feeding in more intensive systems of production. These could include the various chemical pre-treatments, supplementation and the use of multi-nutrient block (Kunju, 1986). Processing (chopping and grinding) of some crop residues may be feasible in certain situations and, associated with this, the development of complete feeds (Reddy, 1987).

An overriding consideration that will determine the value of these approaches is the demonstration of economic benefits. For successful application and acceptance at the farm level, practical technologies need to be simple, within the limits of the farmers' capacity and resource availability, convincing and consistently reproducible.

The need for more evaluative work especially regarding the greater use of fibrous crop residues. Onfarm animal research (OFAR) is probably the only accurate assessment of whether new technology packages are acceptable both economically and socially to farmers, since the technique takes into account the interacting components within farming systems.

OFAR is also a means of identifying and addressing the constraints to adoption of new feeding systems and the extent to which they contribute to sustainability (Devendra, 1990b). Two phases are involved:

Phase 1: Information needed by farmers and advisers:

- Choice of feeds,
- extent of availability and seasonal supply,
- nutritive value
- detailed information on constituents
- prediction of change during production
- capacity to meet production target, and
- capability of filling any nutritional gap
- strategic supplementation:
- on animal performance, and
- on sustainable production of the basal feed resource
- feeding system,
- level and type of production, and
- realistic production targets.

Phase 2: Mechanisms for delivering information to farmers:

- Methodology, a balance between basic and applied research,
- linkages between disciplines and institutions,
- demonstration of economic benefit,
- large-scale on-farm testing,
- *in situ* utilisation
- farmer participation, and
- definition of a model for feed resource development.

Several criteria need to be considered regarding on-farm work and these include *inter alia* the following: production objectives, the treatments involved, the precise methodologies to be used, measurements to be undertaken, type and value of the inputs used and the output derived from the experiment, extent of farmer participation, issues related to the economic analysis of the results and marketing.

Chapter 4

Technologies for Livestock Production

Growing human population, rising per capita income, and increasing urbanization are fuelling rapid growth in the demand for food of animal origin in the developing countries (Delgado et al. 1999). However, current per capita consumption is low. For instance, in India in 1993–94, per capita annual consumption of milk was 51 kg and meat 1.7 kg, much less than the world average of 75 kg milk and 34 kg meat. By 2020 the per capita consumption of milk is likely to more than double and that of meat more than triple.

On the supply side, production of both milk and meat has increased at a rate of about 5% annum. If these trends continue, increases in demand would be met adequately from domestic supplies. The sustainability of these trends, however, is uncertain. In the past, expanding livestock populations mainly contributed to the observed increases in production. India has a huge livestock population comprising different species, and further increase in the livestock population would be constrained severely by the declining land availability. Productivity of Indian livestock is low compared to many developed and developing countries. Cattle milk yield in India is about 12–15% that in the USA, Canada, and Israel. Meat yield of sheep and goats is about 60% less. Feed and fodder scarcity has been the main constraint in raising livestock productivity. Most of the feed requirement is met from crop by-products and grazing on common lands. The latter, however, have been dwindling quantitatively as well as qualitatively. The levels of adoption of breeding-, feed and nutrition-, and health-related technologies are low. It is imperative to raise these since future growth in the livestock subsect or has to come from technological changes.

The importance of livestock in India goes beyond the function of food production. It is an important source of draught power, manure for crop production and fuel for domestic use. Thus, by minimizing use of nonrenewable energy, livestock make a positive contribution to the environment. Although crops and livestock are interdependent to a large extent, the latter constitute an important mechanism for coping with the risks of crop failure. In land-scarce economies livestock provide livelihood support in terms of income and employment generation to the millions of landless and small land holders. In India, livestock wealth is mainly concentrated among the majority of marginal and small land holders. Technology-induced growth in the livestock subsect or would thus improve food and nutritional security, alleviate poverty, and reduce inter regional and interpersonal economic inequities. India spends only about 0.5% of the agricultural gross domestic product on agricultural research. This is low compared to the average of developing countries (0.7%) and developed countries (2.5%). Livestock research receives about 20% of the total agricultural research resources. This corresponds to the contribution of livestock to the agricultural gross domestic product.

Despite the low intensity of investment in research, animal science research over the last few decades has generated a number of technologies in the areas of animal genetics and breeding, feed and nutrition, health, and management. The technical feasibility of many of these has been proven under experimental conditions. Examples include crossbreeding in cattle, sheep, pigs, and poultry; chemical and biological treatment of cereal straws; and vaccines again strinderpest, influenza, and foot and mouth disease.

Studies on returns to investment in livestock research are limited. However, sporadic evidences indicate a very high payoff to investment in livestock research and development. Despite this, the application of many technologies in the field remains limited. Except for crossbreeding, not much information is available regarding adoption and impact of other technologies. There is, thus, considerable scope to raise the productivity of livestock through application of the existing technologies.

Adoption of Technologies

Over the past few decades, animal science research has offered a number of technological options that could raise the productivity of different livestock species if adopted area-wide. These include genetic enhancement of indigenous breeds through crossbreeding with exotic breeds, improvement of nutritive quality of feed and fodder through biological and chemical treatments, development of vaccines against animal diseases, improved livestock management practices, and post harvest management. Additionally, processing technologies have been developed to strengthen the vertical linkages between the farm and dairy industry. The adoption pattern of these technologies varies widely across species, farm typologies and regions.

Breeding Technologies

Genetics and breeding research have evolved many new breeds of cattle, pig, sheep, and poultry using crossbreeding techniques. These breeds have better production coefficients compared to indigenous ones. However, their adoption in the field is limited and sporadic; only about 8% of cattle, 5% of sheep, 15% of pigs, and 33% of poultry populations belong to the crossbred/improved category. The adoption level is higher in urban areas compared to rural areas. In general wide scale adoption of crossbreeds is restricted due to their non-acclimatization to the tropical climates prevailing in most parts of the country.

Besides, their higher maintenance cost, lower disease resistance, and the poor success of artificial insemination (AI) are other barriers to adoption of crossbreeding. There are also species-specific constraints. The crossbred cow has to be replaced frequently to maintain the flow of benefits. Thus, frequent and higher acquisition costs, lack of disposal facilities (cattle slaughter is banned in most Indian states), and poor draught characteristics of male cattle are important impediments to wide-scale adoption of cross breeding technology in cattle. Nonetheless, under certain ecological and economic conditions, adoption of crossbreeding technology in cattle has been quite encouraging. The states of Kerala and Punjab, for instance, have a considerably higher proportion of crossbred cattle.

Sheep husbandry in India is practiced largely by the poor and is dependent on availability of grazing lands, which have been deteriorating in terms of both quantity and quality. In contrast to the indigenous sheep breeds that are capable of surviving even on sparse vegetation, the crossbred sheep require better nutrition. This also applies to pigs, which are often managed in scavenger systems.

In the dairy subsect or the buffalo could emerge as a promising alternative to crossbred cattle because of its adaptability to varied ecological conditions, higher milk yield and higher fat content, realizing a premium price. Unlike cattle, buffalo slaughter is not banned and animals have a disposal value. These factors have favoured a higher growth in buffalo population even without improved breeding interventions. However, in certain regions, nondescript low-yielding species have been upgraded using high-yielding breeds such as the Murrah.

The poultry subsect or has responded well to technological changes and has grown faster than the dairy and ruminant meat sector. Enhancement of genetic potential has been the most important factor in the growth of this sector. However, this has been complemented by health and nutrition technologies. The growth trends are more prominent in specialized periurban/urban poultry systems, because of higher demand for poultry meat and eggs in urban areas. Here, the entry of the private sector has boosted the adoption of technology and growth in poultry outputs. Backyard poultry production, however, continues to languish technologically.

Feed and Nutrition Technologies

A large number of India's livestock, particularly in the arid and semi-arid environments, suffer from inadequate feeding. The feed and fodder shortages, in fact, have been the main limiting factors in raising livestock productivity. Cereal crop residues comprise the main feed for livestock. However, these are deficient in crude protein and several other nutrients. Concentrate feeding is restricted to lactating, high-yielding bovines and work animals. Small ruminants derive their feed requirements mainly from grazing on common lands.

Animal nutrition and crop breeding (straw/stover) research has yielded many new technologies that could augment production and improve the nutritional quality of feeds and fodder. Research on breeding for higher yield and superior quality crop residues (such as in rice, wheat, sorghum, and millets) is in progress. Studies have indicated that a 1% increase in digestibility of sorghum/millet straw increases bovine milk yield by 5–6% (Kristjanson etal. 1998).

Apart from traditional techniques of fodder chopping and conservation, technologies such as urea treatment of fodder, strategic supplementation, urea molasses mineral blocks, and bypass protein use have the potential to alleviate feed and fodder scarcity. These technologies improve digestibility and palatability of feed, reduce feed requirements, avoid feed wastage, and contribute towards improving animal productivity. Some of these techniques, such as fodder chopping and bypass protein use, have long been in practice in many parts of the country, but are not practiced widely. The main constraint to large-scale adoption of nutrition technologies in general has been the lack of information to users.

The area under green fodder crops is also low; constituting no more than 5% of the gross cropped area. The growth in area under fodder crops has been sluggish in most parts of the country, except in the irrigated regions. This is are flection of the rising competition between food and fodder crops for limited land and other resources. Crop breeding research has evolved high-yielding varieties of a number of forage crops. However, these have not been adopted widely due to lack of awareness about new cultivars, nonavailability of irrigation water throughout the year, and problems of insect pests/diseases. Common grazing lands comprise an important source of grasses, and there exists considerable scope to raise the production of grasses/shrubs from these lands through technological and management interventions. Technological interventions such as reseeding with high-yielding grasses and watershed development, complemented with appropriate management interventions such as preventing encroachment, promotion of rotational grazing practices, and charging grazing fees have helped raise the productivity of common grazing resources and thereby improved animal performance.

Disease Control Technologies

Diseases reduce the production potential of livestock. In India many deadly diseases such as rinderpest, foot and mouth disease, hemorrhagic septicemia, and black quarter are major threats to profitable livestock production. Livestock disease control has undergone a paradigm shift in recent years. A number of biological products (vaccines) have been developed for preventive and curative disease management. The infrastructure for disease control has also expanded considerably. The main limitations to effective livestock health management are an inadequate focus on preventive measures, lack of medicines and equipment in the veterinary clinics, and ignorance among the farmers about the diseases and preventive measures. This is reflected in the frequent occurrence of many of these diseases in most parts of the country.

Processing Technologies

Post harvest technologies help producers to realize better gains from technological changes in the primary production sector. But the post harvest processing facilities are lacking in the country. Only about 20% of the total milk production is processed into value-added products.

The bulk of it, however, *is processed into* ghee *and curds by the producers, and a large proportion* of this is consumed at source. Although considerable efforts have gone into developing infrastructure for milk processing in the cooperative sector, only about 5% of the total milk output is processed into table butter, cheese, milk powder, and baby foods.

Information on the proportion of other livestock products entering into the value-added chain is not available. There is a considerable demand for processed meat products, but it remains constricted due to inadequate processing facilities. The same applies to export of these products. Furthermore, not enough attention has been paid towards sanitary and phytosanitary measures. Slaughterhouses are often ill equipped and unhygienic.

Low-cost processing technologies have been developed for both cottage and large industries. The rising demand for processed milk products in urban and rural areas is expected to boost future adoption of processing technologies.

Impact of Technological Change

Technological change improves the production potential of livestock and is reflected in productivity growth. The improvements in production lead to increased welfare of the producers as well as consumers of livestock products, in India, although the intensity of adoption of different technologies is low, the technological changes as discussed above, together with improved management practices have contributed to the increased output of many livestock products.

Productivity and Production

In the last three decades milk production has increased at an annual rate of about 4.5%, and meat and egg production at about 5.5% each. The Total Factor Productivity (TFP) index, which is a joint measure of contribution of technology and technical efficiency, has grown at a rate of 1.4% a year since 1970, while the pre-1970 growth in TFP index was marginally negative.

The growth in TFP is largely a result of yield-augmenting technological changes that have taken place in the dairy and poultry subsectors. The milk yield of cattle and buffalo has grown at a rate of 3.2% and 1.9%, respectively. In the case of poultry the egg yield has almost doubled and the feed conversion efficiency in broiler production has improved tremendously. The growth in productivity of species such as sheep and goats has been negligible.

Consumption, Prices, and Trade

Per capita consumption of livestock products has increased in the last three decades. The share of livestock products in food expenditure has almost doubled to 21% since 1972–73. This is due partly to increases in the availability of livestock products and a decline in prices of major products such as milk and eggs. The real prices of milk, eggs, and pork have declined, partly due to technology-driven growth in outputs. The real prices of products of sheep and goat, where technology uptake had been lacking, witnessed an upward movement.

Improvements in productivity of dairy animals have also helped achieve self-sufficiency in milk production. The imports of milk and milk products have declined to almost zero. Also, exports of certain livestock products are increasing.

Poverty and Equity

Livestock are an important source of income for the rural poor. The growth in the livestock subsect or is expected to contribute to poverty alleviation, as the livestock wealth is largely concentrated among the marginal and small land holders. These categories of farmers, however, face the problem of feed and fodder scarcity. Technologies, particularly those related to nutrition and health, are not capital intensive and could easily be adopted by them. Nevertheless, the intensity of adoption of capital-intensive technologies such as crossbreeding cattle has been observed to be higher on the landless and marginal farms; in 1991–92 there were about 10% crossbreeds in the adult female cattle herds of large land holders, while the landless and marginal land holders maintained 20% of the crossbreeds in their herds. It is thus conjectured that technological change in the livestock subsector would generate more income and employment opportunities for the resource-poor households and contribute towards alleviation of poverty and improvements in interpersonal income distribution.

Rural-Urban Disparities

Peri-urban/urban livestock systems have long existed to cater to the urban demand for foods of animal origin. Rising urbanization and a higher growth in per capita income of the urban population is causing rapid growth in demand for food of animal origin. Developments in processing and packaging technologies have also contributed considerably to this. Thus, to meet the rapidly growing demand for milk, meat, and eggs, peri-urban livestock systems have developed much faster than the rural systems. Technology adoption and production coefficients are better in peri-urban systems. Thus this type of technological dualism is likely to strengthen with further increase in the urban demand for animal foods.

Evidence and Issues

The contribution of science and technology to the growth and development of India's agricultural sector since the mid-1960s is self-evident. Developments in research and production infrastructure and encouraging government policies acted as catalysts for the technology-driven growth in agriculture, helping the country achieve self-sufficiency in food grains and many other commodities. In other areas such as horticulture, fisheries, and livestock considerable technical progress has taken place, but impacts have been slow and sporadic, and its analysis in an economic framework has largely remained undocumented.

The contribution of livestock to income and employment generation is second only to that of crops. However, its productivity is low compared to the world average. Cattle milk yield is about half the world average of 2072 kgannum-1, and the same applies to beef and pork production. The situation becomes more depressing when compared with the developed countries. Average milk yield of cattle is 13% that of the USA and 16% that of Canada. Sheep and pork yields are about 60% less. These figures suggest considerable scope for improvement of India's livestock productivity.

Nonetheless, in recent years certain outputs of livestock such as milk, meat, and eggs have been growing at an annual rate of 5% or more. The sustainability of these trends in the long run is unclear, as the current production environment has several constraints. Feed has been a limiting constraint to ruminant livestock production and this problem is likely to continue in the near future, considering the huge dimensions of livestock populations. Common grazing lands have been deteriorating quantitatively as well qualitatively (Jodha 1991). Animals are fed on crop by-products and grasses from road sides and other marginal lands. Feeding of grains and other concentrates is inadequate, and the competition for grains will intensify with increasing human and livestock populations.

Status of Technological Change

Significant research advances have been made in the areas of animal breeding, nutrition, and health. Many research products have been found to be technically and economically viable under controlled experimental conditions, but the extent of their on-farm application has been rather low. Examples of such technologies include improved animal breeds, and chemical/biological treatment of straw and fodder. A brief review of the existing, as well as potential, technologies that could influence the growth of the livestock subsector is presented below.

Breed Improvement

Crossbreeding of low-yielding indigenous breeds with high-yielding exotic breeds has been widely acknowledged as a valuable strategy to improve animal productivity. Sporadic attempts to improve the genetic potential of different livestock species (poultry, sheep, and cattle) in India were initiated during the latter half of the nineteenth century, but no significant achievements were reported. In the 1950s, systematic crossbreeding research and development programs were initiated. Crossbreeding research has focused mainly on cattle because of their dual role of milk production and use as draught animals in the crop subsector. Between 1982 and 1992, the population of crossbred cattle increased at an annual rate of 5.6%. On the other hand, indigenous stock seemed to be approaching stabilization. The annual growth rate of indigenous cattle was 0.5% although the female population witnessed a slightly higher growth. Low milk yield and decreasing demand for draught animals are the main factors for slow growth in indigenous stock. These trends indicate a gradual substitution of indigenous cattle by crossbreds.

Despite being an important milch species, the buffalo has not received much attention in breed improvement. Development efforts have focused on upgrading low-yielding breeds through artificial insemination. The buffalo population increased at a rate of 1.9% per year between 1982 and 1992. The female population witnessed faster growth than the male population. Its adaptability to a wide range of climatic conditions, higher milk yield compared to indigenous cattle, and price premium on milk due to its higher fat content have favoured faster growth in the buffalo population. Furthermore, the disposal value of buffalo is higher; unlike cattle there are fewer restrictions on buffalo slaughter.

The population of improved poultry grew at an annual rate of about 9%, more than double that of indigenous poultry. Technological transformation of the poultry subsector seems to be market-driven as the demand for poultry meat and eggs is income-elastic and has been rising continuously. The populations of crossbred sheep and pigs also grew faster than their indigenous counterparts.

Empirical evidence from the field proves the scientific claims of better economic performance of crossbred animals. Despite this, cross breeding technology has not gained a foothold. One of the possible reasons is non acclimatization of crossbred animals to varying climatic conditions in the country, causing a number of health-and physiology-related problems. In this context, Steane (1999) mentions that 'in using exotic breeds as a strategy for improvement it seems to have been assumed that genotype-environmental interactions do not exist or the optimal economic and sustainable crossing structure would automatically be developed. Indeed in much of Asia, there appears to be a history of introducing breeds without proper evaluation and with little or no thought given to the breeding structure which will best use the available material'.

Higher initial investment and maintenance costs also limit wide spread adoption of crossbreds. The first cross animals perform very well, but the performance of animals from subsequent crosses declines significantly. Therefore crossbred animals need to be replaced frequently in order to sustain the flow of benefits. Frequent acquisition of first crossbreds, without realizing appropriate disposal value of the subsequently crossed animals, renders cross breeding technology capital intensive. In this context, Alderman (1987) observed that more than 50% of the farmers in Karnataka maintaining crossbreeds depend on the market for replacement of first cross animals. This is to avoid the risk of getting unwanted male calves and the associated problems of breeding and feeding the calf.

As indigenous cattle are sources of both milk and draught power for agriculture, their replacement by crossbred cattle has been slow. Cross bred males are considered inefficient for draught purposes compared to indigenous males, although animal physiology research has shown that the difference is marginal. Furthermore, machines have emerged as a major source of power in Indian agriculture, although mechanization does not appear to have affected the population of working animals significantly. The number of tractors per thousand ha increased from 0.6 in 1972 to 10.9 in 1992, while the working bovines declined marginally.

Mechanization has affected working bovines mainly on medium and large farms. The number of working bovines declined from 410 in 1972 to 180 per 1000 ha in 1992 on medium farms and from 210 to 50 on large farms. On small and marginal farms, most of which did not have tractors, working animals provided most of the draught power requirement in 1972. In 1992, the number of tractors on these farms had increased to 4.1 and 7.8 per thousand ha without a concomitant decline in the number of working animals.

Nonetheless, crossbreeding strategies have been successful under certain environments and economic conditions as in Kerala and Punjab. In Kerala the milk production system is cattle-based. Indigenous breeds are poor as milk producers and for draught power (Rajapurohit 1979). Cropping pattern is largely plantation-oriented, requiring less draught power. Furthermore, unlikein many other states, cattle slaughter is not prohibited by law in Kerala, thus making it easier to cull out the low-yielding and unwanted animals. In Punjab, on the other hand, increasing intensification of agriculture required more power to perform agricultural operations on time, which indigenous cattle were not capable of providing. Moreover, feed resources have never been a problem in Punjab.

Feed and Nutrition

Adequate provision of feed is essential to livestock production, and its scarcity has been one of the major limiting factors in improving productivity in India. Crop residues and by-products comprise the main feeds, accounting for 40% of the total consumption (World Bank 1996). Green fodder contributes 26%, while concentrates contribute only 3 percent. The rest comes from grazing. Stall-feeding is largely confined to buffalo, crossbred cattle, and draught animals. Small ruminants mainly feed on grasses from village common property resources, road sides, and harvested fields, and supplementary feeding is lacking except in large commercial herds.

Role of Women in Livestock Farming

As farmers, women in subsistence production ensure the survival of millions of people in all regions. Women in sustenance economies, producing and reproducing wealth in partnership with nature, are experts in their own right with ecological knowledge of nature's processes. Women's livelihood strategies, and their support and means of food security are diverse and complex, from cultivating field crops to livestock rearing, to home gardening, gathering, fishing from sources such as swamps, forests, woodlands, wastelands and quarries, etc. But these alternative modes of knowledge and livelihoods are not recognised by conventional agricultural scientists and development experts, who fail to see the connection of women's knowledge, work and skills with ensuring community food security, and the creation of wealth.

More than half of the world's food is grown by women. Women's work is both wide-ranging and multifaceted throughout the year, and they perform multiple tasks in the sphere of agriculture. Women's indigenous knowledge and skills are vitally necessary for food production and sustainable agriculture. Women's intimate knowledge of seed preparation and soil management, plants and pest control, post-harvest processing and storage, animal husbandry, as well as food processing and meal preparation are significant-crucial also to ensuring food security through sustainable agriculture. They are the authority on the interface of livestock keeping with farming.

However, there is little recognition of their significant role and contribution to the socio-economic development of a nation. The entrenched social and religious norms that define women's role as secondary and subordinate keep women vulnerable and dependent and allow women's exploitation as agricultural workers and farmers. Ensuring that the majority of rural women do not do not own land nor have access to productive resources.

As a result of rapid globalization, this patriarchal construct of women's role, position in society and the gender division of labour, in terms of production and reproduction, continue to operate. For example, women in rural agriculture face loss of livelihoods with the dumping of subsidised imported food products into her country. As a result, women who traditionally have the gendered responsibility of household food security are forced to work longer hours, and even harder to ensure the food security of the family. Women are faced with increased costs of food production as inputs costs rise, and women's health is increasingly compromised from exposure to hazardous pesticides. All these factors have intensified rural women's impoverishment, displacement and hunger. Another example, during the East Asian economic crisis, women again had to put more effort in trying to make up for lost urban incomes. In the absence of state-schemes of income and employment support, it was largely left to the so-called coping strategies of women to deal with the rural effects of the Asian crisis. Globalization in agriculture has further intensified the development of the corporate agriculture and contract farming that perpetuates monocultures and high input agriculture. It will also intensify the control of agriculture production and distribution, and inputs into the hands of a few Transnational Corporations (TNCs) with headquarters in the North.

Women are also exploited as workers in the corporate farms that are expanding into rural areas. In these large farms or plantations, it is the women worker who is sought out and hired. Employers see women agricultural workers, as unskilled workers who will accept low wages and increased workloads without complaining and women are known to rarely join unions or organise for their rights. Women are more exposed to pesticides that adversely impact their health and their family and community's health.

This system of agriculture undoubtedly marginalize women's knowledge and skills with the introduction of new philosophies and technologies-thus eroding the base of whatever little power they had traditionally. For women it means loss of control over their knowledge and resources including seeds, plants, herbs and other devices for pest management, and commons and forest land for food and fodder gathering. Such knowledge will either become obsolete in the age of corporate agriculture or "stolen" for commercial distribution by TNCs. Productive resources will be corporatized in the name of development and globalization.

Environmental Impacts of Livestock Production

Like many industries, Industrial Farm Animal Production (ifap) results in a number of environmental impacts that affect populations both near and far. While every industry may contribute to society via production of some necessary or desired good, as our population increases, we have become more and more aware of the finite nature of our world's resources and of the impacts of our various industries upon those resources and our own human health. Industrial farm operations impact all major environmental media, including water, soil, and air. Of most concern are the pollution of ground and surface water resources with nutrients, industrial and agricultural chemicals, and microorganisms; the use of freshwater resources; the contamination and degradation of soil; and the release of toxic gases and odorous substances, as well as particulates and bioaerosols containing microorganisms and pathogens. The Commission queried the authors of this report on the magnitude and key determinants of these impacts, and the resulting impacts on both human health and ecosystems.

The major causes of the above noted environmental impacts of ifap are the enormous amounts of waste that are produced in a very small area in this agricultural model, the inadequate systems we now have to deal with that waste, and the large energy and resource inputs required for this type of production, including feed production and transport. The usda Agricultural Research Services (ars) estimated the manure output from farm animals in the United States to be nearly 1 million US short tons of dry matter per day in 2001. Eighty-six percent of this was estimated to be produced by animals held in confinement. Different groups have posited both lower and higher estimates, but the fact remains that food animals produce an enormous amount of waste every day, exceeding human sanitary waste production by at least one order of magnitude. However, disposal of this waste is far less closely regulated than disposal of human waste. Animal manure and other agricultural waste result in water and air degradation, which in turn impact both the aquatic and the terrestrial ecosystems surrounding these operations.

In addition to the enormous waste produced by industrial agriculture, this system requires major inputs of both energy and resources. Water use is more significant in these systems because it is often used for cleaning the buildings and in the waste management systems. In addition, the industrial model utilizes feed, which is grown in monocultures, often far away from the facility. Enormous quantities of both water and petroleum-based pesticides may be used in the production of this feed, leading not only to the depletion of water resources, but also to soil erosion and pollution with pesticides. Pesticide residues may remain in the animal feed, leading to the possibility of toxic residues in the food animals themselves. Feed crop monocultures also contribute to loss of biodiversity, as they are planted in place of other plants and/or animal habitats.

Finally, but growing more urgent every day, industrial agriculture may be a significant contributor to climate change, as the production of greenhouse gases from these facilities (both from the animals themselves and from the decomposition of their waste) is significant. Taken together, these data suggest that the present industrial model of farm animal production is not sustainable for the long term. The overuse and degradation of natural resources may be too great to allow the current form of this production model to continue to be viable. The commission requested that the authors of this report investigate the scope of these environmental factors, to help grasp the breadth of the possible impacts of the ifap system.

Global meat production has tripled in the past three decades and could double its present level by 2050, according to a new report on the livestock industry by an international team of scientists and policy experts. The impact of this "livestock revolution" is likely to have significant consequences for human health, the environment and the global economy, the authors conclude.

Among the key findings in the report are:

- More than 1.7 billion animals are used in livestock production worldwide and occupy more than one-fourth of the Earth's land.
- Production of animal feed consumes about one-third of total arable land.

- Livestock production accounts for approximately 40 percent of the global agricultural gross domestic product.
- The livestock sector, including feed production and transport, is responsible for about 18 percent of all greenhouse gas emissions worldwide.

Impacts on Humanity

Although about 1 billion poor people worldwide derive at least some part of their livelihood from domesticated animals, the rapid growth of commercialized industrial livestock has reduced employment opportunities for many, according to the report. In developing countries, such as India and China, large-scale industrial production has displaced many small, rural producers, who are under additional pressure from health authorities to meet the food safety standards that a globalized marketplace requires.

Beef, poultry, pork and other meat products provide one-third of humanity's protein intake, but the impact on nutrition across the globe is highly variable, according to the report. "Too much animal-based protein is not good for human diets, while too little is a problem for those on a protein-starved diet, as happens in many developing countries," Mooney noted. While over consumption of animal-source foods – particularly meat, milk and eggs – has been linked to heart disease and other chronic conditions, these foods remain a vital source of protein and nutrient nutrition throughout the developing world, the report said. The authors cited a recent study of Kenyan children that found a positive association between meat intake and physical growth, cognitive function and school performance.

Human health also is affected by pathogens and harmful substances transmitted by livestock, the authors said. Emerging diseases, such as highly pathogenic avian influenza, are closely linked to changes in the livestock production but are more difficult to trace and combat in the newly globalized marketplace, they said.

Environmental Impacts

The livestock sector is a major environmental polluter, the authors said, noting that much of the world's pasture land has been degraded by grazing or feed production, and that many forests have been clear-cut to make way for additional farmland. Feed production also requires intensive use of water, fertilizer, pesticides and fossil fuels, added co-editor Henning Steinfeld of the United Nations Food and Agriculture Organization (FAO).

A new report from FAO says livestock production is one of the major causes of the world's most pressing environmental problems, including global warming, land degradation, air and water pollution, and loss of biodiversity. Using a methodology that considers the entire commodity chain, it estimates that livestock are responsible for 18 percent of greenhouse gas emissions, a bigger share than that of transport. However, the report says, the livestock sector's potential contribution to solving environmental problems is equally large, and major improvements could be achieved at reasonable cost.

Based on the most recent data available, Livestock's long shadow takes into account the livestock sector's direct impacts, plus the environmental effects of related land use changes and production of the feed crops animals consume. It finds that expanding population and incomes worldwide, along with changing food preferences, are stimulating a rapid increase in demand for meat, milk and eggs, while globalization is boosting trade in both inputs and outputs.

In the process, the livestock sector is undergoing a complex process of technical and geographical change. Production is shifting from the countryside to urban and peri-urban areas, and towards sources of animal feed, whether feed crop areas or transport and trade hubs where feed is distributed. There is also a shift in species, with accelerating growth in production of pigs and poultry (mostly in industrial units) and a slow-down in that of cattle, sheep and goats, which are often raised extensively. Today, an estimated 80 percent of growth in the livestock sector comes from industrial production systems. Owing to those shifts, the report says, livestock are entering into direct competition for scarce land, water and other natural resources.

Deforestation, Greenhouse Gases

The livestock sector is by far the single largest anthropogenic user of land. Grazing occupies 26 percent of the Earth's terrestrial surface, while feed crop production requires about a third of all arable land. Expansion of grazing land for livestock is a key factor in deforestation, especially in Latin America: some 70 percent of previously forested land in the Amazon is used as pasture, and feed crops cover a large part of the reminder. About 70 percent of all grazing land in dry areas is considered degraded, mostly because of overgrazing, compaction and erosion attributable to livestock activity.

At the same time, the livestock sector has assumed an often unrecognized role in global warming. Using a methodology that considered the entire commodity chain, FAO estimated that livestock are responsible for 18 percent of greenhouse gas emissions, a bigger share than that of transport. It accounts for nine percent of anthropogenic carbon dioxide emissions, most of it due to expansion of pastures and arable land for feed crops. It generates even bigger shares of emissions of other gases with greater potential to warm the atmosphere: as much as 37 percent of anthropogenic methane, mostly from enteric fermentation by ruminants, and 65 percent of anthropogenic nitrous oxide, mostly from manure.

Livestock production also impacts heavily the world's water supply, accounting for more than 8 percent of global human water use, mainly for the irrigation of feed crops. Evidence suggests it is the largest sectoral source of water pollutants, principally animal wastes, antibiotics, hormones, chemicals from tanneries, fertilizers and pesticides used for feed crops, and sediments from eroded pastures. While global figures are unavailable, it is estimated that in the USA livestock and feed crop agriculture are responsible for 37 percent of pesticide use, 50 percent of antibiotic use, and a third of the nitrogen and phosphorus loads in freshwater resources. The sector also generates almost two-thirds of anthropogenic ammonia, which contributes significantly to acid rain and acidification of ecosystems.

The sheer quantity of animals being raised for human consumption also poses a threat of the Earth's biodiversity. Livestock account for about 20 percent of the total terrestrial animal biomass, and the land area they now occupy was once habitat for wildlife.

In 306 of the 825 terrestrial eco-regions identified by the Worldwide Fund for Nature, livestock are identified as "a current threat", while 23 of Conservation International's 35 "global hotspots for biodiversity"-characterized by serious levels of habitat loss-are affected by livestock production. In particular, water is grossly under-priced in most countries, and development of water markets and various types of cost recovery will be needed to correct the situation. In the case of land, suggested instruments include grazing fees, and better institutional arrangements for controlled and equitable access. The removal of livestock production subsidies is also likely to improve technical efficiency-in New Zealand, a drastic reduction in agricultural subsidies during the 1980s helped create one of the world's most efficient and environmentally friendly ruminant livestock industries.

Removal of price distortions at input and product level will enhance natural resource use, but may often not be sufficient. Livestock's long shadow says environmental externalities, both negative and positive, need to be explicitly factored into the policy framework. Livestock holders who provide environmental services need to be compensated, either by the immediate beneficiary (such as downstream users enjoying improved water quantity and quality) or by the general public. Services that could be rewarded include land management or land uses that restore biodiversity, and pasture management that provides for carbon sequestration. Compensation schemes also need to be developed between water and electricity providers and graziers who adopt grass lands management strategies that reduce sedimentation of water reservoirs. Likewise, livestock holders who emit waste into waterways or release ammonia into the atmosphere should pay for the damage. Applying the "polluter pays" principle should not present insurmountable problems for offenders, given the burgeoning demand for livestock products.

Consumer Pressure

Finally, FAO says, the livestock sector is usually driven by diverse policy objectives, and decision-makers find it difficult to address economic, social, health and environmental issues at the same time. The fact that so many people depend on livestock for their livelihoods limits the policy options available, and leads to difficult and politically sensitive trade-offs. Information, communication and education will play critical roles in enhancing a "willingness to act". With their strong and growing influence, consumers are likely to be the main source of commercial and political pressure "to push the livestock sector into more sustainable forms", Livestock's long shadow says.

Already, growing awareness of threats to the environment is translating into rising demand for environmental services: "This demand will broaden from immediate concerns-such as reducing the nuisance of flies and odours-to intermediate demands for clean air and water, then to the broader, longer-term environmental concerns, including climate change and loss of biodiversity".

Applications of Animal Biotechnology

Animal biotechnology is the application of scientific and engineering principles to the processing or production of materials by animals or aquatic species to provide goods and services (NRC 2003). Examples of animal biotechnology include generation of transgenic animals or transgenic fish (animals or fish with one or more genes introduced by human intervention), using gene knockout technology to generate animals in which a specific gene has been inactivated, production of nearly identical animals by somatic cell nuclear transfer (also referred to as clones), or production of infertile aquatic species.

Transgenics

Since the early 1980s, methods have been developed and refined to generate transgenic animals or transgenic aquatic species. For example, transgenic livestock and transgenic aquatic species have been generated with increased growth rates, enhanced lean muscle mass, enhanced resistance to disease or improved use of dietary phosphorous to lessen the environmental impacts of animal manure.

Transgenic poultry, swine, goats, and cattle also have been produced that generate large quantities of human proteins in eggs, milk, blood, or urine, with the goal of using these products as human pharmaceuticals. Examples of human pharmaceutical proteins include enzymes, clotting factors, albumin, and antibodies. The major factor limiting widespread use of transgenic animals in agricultural production systems is the relatively inefficient rate (success rate less than 10 percent) of production of transgenic animals.

Gene Knockout Technology

Animal biotechnology also can knock out or inactivate a specific gene. Knockout technology creates a possible source of replacement organs for humans. The process of transplanting cells, tissues, or organs from one species to another is referred to as "xenotransplantation." Currently, the pig is the major animal being considered as a xenotransplant donor to humans. Unfortunately, pig cells and human cells are not immunologically compatible. Pig cells express a carbohydrate epitope (alpha1, 3 galactose) on their surface that is not normally found on human cells. Humans will generate antibodies to this epitope, which will result in acute rejection of the xenograft. Genetic engineering is used to knock out or inactivate the pig gene (alpha1, 3 galactosyl transferase) that attaches this carbohydrate epitope on pig cells. Other examples of knockout technology in animals include inactivation of the prion-related peptide (PRP) gene that may generate animals resistant to diseases associated with prions (bovine spongiform encephalopathy [BSE], Creutzfeldt-Jakob Disease [CJD], scrapie, etc.). Most of the funding for these types of projects is conducted by private companies or in academic laboratories supported by the National Institutes of Health. Research projects designed to provide basic information regarding mechanisms associated with gene knockout technology are supported by NIFA.

Somatic Cell Nuclear Transfer

Another application of animal biotechnology is the use of somatic cell nuclear transfer to produce multiple copies of animals that are nearly identical copies of other animals (transgenic animals, genetically superior animals, or animals that produce high quantities of milk or have some other desirable trait, etc.). This process has been referred to as cloning. To date, somatic cell nuclear transfer has been used to clone cattle, sheep, pigs, goats, horses, mules, cats, rats, and mice. The technique involves culturing somatic cells from an appropriate tissue (fibroblasts) from the animal to be cloned. Nuclei from the cultured somatic cells are then microinjected into an enucleated oocyte obtained from another individual of the same or a closely related species.

Through a process that is not yet understood, the nucleus from the somatic cell is reprogrammed to a pattern of gene expression suitable for directing normal development of the embryo. After further culture and development in vitro, the embryos are transferred to a recipient female and ultimately will result in the birth of live offspring. The success rate for propagating animals by nuclear transfer is often less than 10 percent and depends on many factors, including the species, source of the recipient ova, cell type of the donor nuclei, treatment of donor cells prior to nuclear transfer, the techniques employed for nuclear transfer, etc. NIFA has supported research projects to obtain a better understanding of the basic cellular mechanisms associated with nuclear reprogramming.

Production of Infertile Aquatic Species. In aquaculture production systems, some species are not indigenous to a given area and can pose an ecological risk to native species should the foreign species escape confinement and enter the natural ecosystem. Generation of large populations of sterile fish or mollusks is one potential solution to this problem. Techniques have been developed to alter the chromosome complement to render individual fish and mollusks infertile.

For example, triploid individuals (with three, instead of two, sets of chromosomes) have been generated by using various procedures to interfere with the final step in meiosis (extrusion of the second polar body). Timed application of high or low temperatures, various chemicals, or high hydrostatic pressure to newly fertilized eggs has been effective in producing triploid individuals. At a later time, the first cell division of the zygote can be suppressed to produce a fertile tetraploid individual (four sets of chromosomes). Tetraploids can then be mated with normal diploids to produce large numbers of infertile triploids. Unfortunately, in a commercial production system, it is often difficult to obtain sterilization of 100 percent of the individuals; thus, alternative methods are needed to ensure reproductive confinement of transgenic fish. Another technique that is being developed for finfish is to farm monosex fish stocks. Monosex populations can be produced by gender reversal and progeny testing to identify XX males for producing all female stocks or YY males for producing all male stocks.

As with any new technology, animal biotechnology faces a variety of uncertainties, safety issues and potential risks. For example, concerns have been raised regarding: the use of unnecessary genes in constructs used to generate transgenic animals, the use of vectors with the potential to be transferred or to otherwise contribute sequences to other organisms, the potential effects of genetically modified animals on the environment, the effects of the biotechnology on the welfare of the animal, and potential human health and food safety concerns for meat or animal products derived from animal biotechnology. Before animal biotechnology will be used widely by animal agriculture production systems, additional research will be needed to determine if the benefits of animal biotechnology outweigh these potential risks. The USDA Biotechnology Risk Assessment Grants program supports environmental risk assessment research projects on genetically engineered animalsAdvances in animal biotechnology have been facilitated by recent progress in sequencing and analyzing animal genomes, identification of molecular markers (microsatellites, expressed sequence tags [ESTs], quantitative trait loci [QTLs], etc.) and a better understanding of the mechanisms that regulate gene expression.

A Long History

The animal biotechnology in use today is built on a long history. Some of the first biotechnology in use includes traditional breeding techniques that date back to 5000 B.C.E. Such techniques include crossing diverse strains of animals (known as hybridizing) to produce greater genetic variety. The offspring from these crosses then are bred selectively to produce the greatest number of desirable traits. For example, female horses have been bred with male donkeys to produce mules, and male horses have been bred with female donkeys to produce hinnies, for use as work animals, for the past 3,000 years. This method continues to be used today.

The modern era of biotechnology began in 1953, when American biochemist James Watson and British biophysicist Francis Crick presented their double-helix model of DNA. That was followed by Swiss microbiologist Werner Arber's discovery in the 1960s of special enzymes, called restriction enzymes, in bacteria. These enzymes cut the DNA strands of any organism at precise points. In 1973, American geneticist Stanley Cohen and American biochemist Herbert Boyer removed a specific gene from one bacterium and inserted it into another using restriction enzymes. That event marked the beginning of recombinant DNA technology, or genetic engineering. In 1977, genes from other organisms were transferred to bacteria, an achievement that led eventually to the first transfer of a human gene.

The Technology Involved

Other Technologies

Animal biotechnology in use today is based on the science of genetic engineering. Under the umbrella of genetic engineering exist other technologies, such as transgenics and cloning, that also are used in animal biotechnology. In addition to the use of transgenics and cloning, scientists can use gene knock out technology to inactivate, or "knock out," a specific gene. It is this technology that creates a possible source of replacement organs for humans. The process of transplanting cells, tissues or organs from one species to another is referred to as xenotransplantation. Currently, the pig is the major animal being considered as a viable organ donor to humans. Unfortunately, pig cells and human cells are not immunologically compatible. Pigs, like almost all mammals, have markers on their cells that enable the human immune system to recognize them as foreign and reject them. Genetic engineering is used to knock out the pig gene responsible for the protein that forms the marker to the pig cells.

Cloning

Scientists use reproductive cloning techniques to produce multiple copies of mammals that are nearly identical copies of other animals, including transgenic animals, genetically superior animals and animals that produce high quantities of milk or have some other desirable trait.

To date, cattle, sheep, pigs, goats, horses, mules, cats, rats and mice have been cloned, beginning with the first cloned animal, a sheep named Dolly, in 1996. Reproductive cloning begins with somatic cell nuclear transfer (SCNT). In SCNT, scientists remove the nucleus from an egg cell (oocyte) and replace it with a nucleus from a donor adult somatic cell, which is any cell in the body except for an oocyte or sperm. For reproductive cloning, the embryo is implanted into a uterus of a surrogate female, where it can develop into a live being.

Its Applications

Animal biotechnology has many potential uses. Since the early 1980s, transgenic animals have been created with increased growth rates, enhanced lean muscle mass, enhanced resistance to disease or improved use of dietary phosphorous to lessen the environmental impacts of animal manure.

Transgenic poultry, swine, goats and cattle that generate large quantities of human proteins in eggs, milk, blood or urine also have been produced, with the goal of using these products as human pharmaceuticals. Human pharmaceutical proteins include enzymes, clotting factors, albumin and antibodies. The major factor limiting the widespread use of transgenic animals in agricultural production systems is their relatively inefficient production rate (a success rate of less than 10 percent).

A specific example of these particular applications of animal biotechnology is the transfer of the growth hormone gene of rainbow trout directly into carp eggs. The resulting transgenic carp produce both carp and rainbow trout growth hormones and grow to be one-third larger than normal carp.

Another example is the use of transgenic animals to clone large quantities of the gene responsible for a cattle growth hormone. The hormone is extracted from the bacterium, is purified and is injected into dairy cows, increasing their milk production by 10 to 15 percent. That growth hormone is called bovine somatotropin or BST.

Another major application of animal biotechnology is the use of animal organs in humans. Pigs currently are used to supply heart valves for insertion into humans, but they also are being considered as a potential solution to the severe shortage in human organs available for transplant procedures.

The Future of Animal Biotechnology

While predicting the future is inherently risky, some things can be said with certainty about the future of animal biotechnology. The government agencies involved in the regulation of animal biotechnology, mainly the Food and Drug Administration (FDA), likely will rule on pending policies and establish processes for the commercial uses of products created through the technology. In fact, as of March 2006, the FDA was expected to rule in the next few months on whether to approve meat and dairy products from cloned animals for sale to the public. If these animals and animal products are approved for human consumption, several companies reportedly are ready to sell milk, and perhaps meat, from cloned animals — most likely cattle and swine. It also is expected that technologies will continue to be developed in the field, with much hope for advances in the use of animal organs in human transplant operations.

Related Issues

The potential benefits of animal biotechnology are numerous and include enhanced nutritional content of food for human consumption; a more abundant, cheaper and varied food supply; agricultural land-use savings; a decrease in the number of animals needed for the food supply; improved health of animals and humans; development of new, low-cost disease treatments for humans; and increased understanding of human disease.

Yet despite these potential benefits, several areas of concern exist around the use of biotechnology in animals. To date, a majority of the American public is uncomfortable with genetic modifications to animals. According to a survey conducted by the Pew Initiative on Food and Biotechnology, 58 percent of those polled said they opposed scientific research on the genetic engineering of animals. And in a Gallup poll conducted in May 2004, 64 percent of Americans polled said they thought it was morally wrong to clone animals. Concerns surrounding the use of animal biotechnology include the unknown potential health effects to humans from food products created by transgenic or cloned animals, the potential effects on the environment and the effects on animal welfare. Before animal biotechnology will be used widely by animal agriculture production systems, additional research will be needed to determine if the benefits of animal biotechnology outweigh these potential risks.

Food Safety

The main question posed about the safety of food produced through animal biotechnology for human consumption is, “Is it safe to eat?” But answering that question isn’t simple. Other questions must be answered first, such as, “What substances expressed as a result of the genetic modification are likely to remain in food?” Despite these questions, the

National Academies of Science (NAS) released a report titled Animal Biotechnology: Science-Based Concernsstating that the overall concern level for food safety was determined to be low. Specifically, the report listed three specific food concerns: allergens, bioactivity and the toxicity of unintended expression products.

The potential for new allergens to be expressed in the process of creating foods from genetically modified animals is a real and valid concern, because the process introduces new proteins. While food allergens are not a new issue, the difficulty comes in how to anticipate these adequately, because they only can be detected once a person is exposed and experiences a reaction. Another food safety issue, bioactivity, asks, "Will putting a functional protein like a growth hormone in an animal affect the person who consumes food from that animal?" As of May 2006, scientists cannot say for sure if the proteins will. Finally, concern exists about the toxicity of unintended expression products in the animal biotechnology process. While the risk is considered low, there is no data available. The NAS report stated it still needs be proven that the nutritional profile does not change in these foods and that no unintended and potentially harmful expression products appear.

Environmental Considerations

Another major concern surrounding the use of animal biotechnology is the potential for negative impact to the environment. These potential harms include the alteration of the ecologic balance regarding feed sources and predators, the introduction of transgenic animals that alter the health of existing animal populations and the disruption of reproduction patterns and their success.

To assess the risk of these environmental harms, many more questions must be answered, such as: What is the possibility the altered animal will enter the environment? Will the animal's introduction change the ecological system? Will the animal become established in the environment? and Will it interact with and affect the success of other animals in the new community? Because of the many uncertainties involved, it is challenging to make an assessment.

To illustrate a potential environmental harm, consider that if transgenic salmon with genes engineered to accelerate growth were released into the natural environment, they could compete more successfully for food and mates than wild salmon. Thus, there also is concern that genetically engineered organisms will escape and reproduce in the natural environment. It is feared existing species could be eliminated, thus upsetting the natural balance of organisms.

Legal Implications

Regulations

The regulation of animal biotechnology currently is performed under existing government agencies. To date, no new regulations or laws have been enacted to deal with animal biotechnology and related issues.

The main governing body for animal biotechnology and their products is the FDA. Specifically, these products fall under the new animal drug provisions of the Food, Drug, and Cosmetic Act (FDCA). In this use, the introduced genetic construct is considered the "drug." This lack of concrete regulatory guidance has produced many questions, especially because the process for bringing genetically engineered animals to market remains unknown.

Currently, the only genetically engineered animal on the market is the GloFish, a transgenic aquarium fish engineered to glow in the dark. It has not been subject to regulation by the FDA, however, because it is not believed to be a threat to the environment.

Many people question the use of an agency that was designed specifically for drugs to regulate live animals. The agency's strict confidentiality provisions and lack of an environmental mandate in the FDCA also are concerns. It still is unclear how the agency's provisions will be interpreted for animals and how multiple agencies will work together in the regulatory system.

When animals are genetically engineered for biomedical research purposes (as pigs are, for example, in organ transplantation studies), their care and use is carefully regulated by the Department of Agriculture. In addition, if federal funds are used to support the research, the work further is regulated by the Public Health Service Policy on

Humane Care and use of Laboratory Animals

Labeling

Whether products generated from genetically engineered animals should be labelled is yet another controversy surrounding animal biotechnology. Those opposed to mandatory labeling say it violates the government's traditional focus on regulating products, not processes..

If a product of animal biotechnology has been proven scientifically by the FDA to be safe for human consumption and the environment and not materially different from similar products produced via conventional means, these individuals say it is unfair and without scientific rationale to single out that product for labeling solely because of the process by which it was made.

On the other hand, those in favour of mandatory labeling argue labeling is a consumer "right-to-know" issue. They say consumers need full information about products in the marketplace — including the processes used to make those products — not for food safety or scientific reasons, but so they can make choices in line with their personal ethics.

Intellectual Property

On average, it takes seven to nine years and an investment of about $55 million to develop, test and market a new genetically engineered product. Consequently, nearly all researchers involved in animal biotechnology are protecting their investments and intellectual property through the patent system. In 1988, the first patent was issued on a transgenic animal, a strain of laboratory mice whose cells were engineered to contain a cancer-predisposing gene. Some people, however, are opposed ethically to the patenting of life forms, because it makes organisms the property of companies. Other people are concerned about its impact on small farmers. Those opposed to using the patent system for animal biotechnology have suggested using breed registries to protect intellectual property.

Ethical and Social Considerations

Ethical and social considerations surrounding animal biotechnology are of significant importance. This especially is true because researchers and developers worry the future market success of any products derived from cloned or genetically engineered animals will depend partly on the public's acceptance of those products.

Animal biotechnology clearly has its skeptics as well as its outright opponents. Strict opponents think there is something fundamentally immoral about the processes of transgenics and cloning. They liken it to "playing God." Moreover, they often oppose animal biotechnology on the grounds that it is unnatural. Its processes, they say, go against nature and, in some cases, cross natural species boundaries. Still others question the need to genetically engineer animals. Some wonder if it is done so companies can increase profits and agricultural production. They believe a compelling need should exist for the genetic modification of animals and that we should not use animals only for our own wants and needs. And yet others believe it is unethical to stifle technology with the potential to save human lives. While the field of ethics presents more questions than it answers, it is clear animal biotechnology creates much discussion and debate among scientists, researchers and the American public. Two main areas of debate focus on the welfare of animals involved and the religious issues related to animal biotechnology.

Animal Welfare

Perhaps the most controversy and debate regarding animal biotechnology surrounds the animals themselves. While it has been noted that animals might, in fact, benefit from the use of animal biotechnology — through improved health, for example — the majority of discussion is about the known and unknown potential negative impacts to animal welfare through the process. For example, calves and lambs produced through in vitro fertilization or cloning tend to have higher birth weights and longer gestation periods, which leads to difficult births that often require cesarean sections.

In addition, some of the biotechnology techniques in use today are extremely inefficient at producing fetuses that survive of the transgenic animals that do survive, many do not express the inserted gene properly, often resulting in anatomical, physiological or behavioural abnormalities. There also is a concern that proteins designed to produce a pharmaceutical product in the animal's milk might find their way to other parts of the animal's body, possibly causing adverse effects. Animal "telos" is a concept derived from Aristotle and refers to an animal's fundamental nature. Disagreement exists as to whether it is ethical to change an animal's telos through transgenesis. For example, is it ethical to create genetically modified chickens that can tolerate living in small cages?

Those opposed to the concept say it is a clear sign we have gone too far in changing that animal. Those unopposed to changing an animal's telos, however, argue it could benefit animals by fitting them for living conditions for which they are not "naturally" suited. In this way, scientists could create animals that feel no pain.

Religious Issues

Religion plays a crucial part in the way some people view animal biotechnology. For some people, these technologies are considered blasphemous. In effect, God has created a perfect, natural order, they say, and it is sinful to try to improve that order by manipulating the basic ingredient of all life, DNA. Some religions place great importance on the "integrity" of species, and as a result, those religion's followers strongly oppose any effort to change animals through genetic modification. Not all religious believers make these assertions, however, and different believers of the same religion might hold differing views on the subject. For example, Christians do not oppose animal biotechnology unanimously. In fact, some Christians support animal biotechnology, saying the Bible teaches humanity's dominion over nature. Some modern theologians even see biotechnology as a challenging, positive opportunity for us to work with God as "co-creators." Transgenic animals can pose problems for some religious groups. For example, Muslims, Sikhs and Hindus are forbidden to eat certain foods. Such religious requirements raise basic questions about the identity of animals and their genetic makeup. If, for example, a small amount of genetic material from a fish is introduced into a melon (in order to allow it grow to in lower temperatures), does that melon become "fishy" in any meaningful sense?

Some would argue all organisms share common genetic material, so the melon would not contain any of the fish's identity. Others, however, believe the transferred genes are exactly what make the animal distinctive; therefore the melon would be forbidden to be eaten as well.

Thinking Critically

The following questions are intended for middle school and high school students studying animal biotechnology. They are designed to stimulate critical thinking about the many ethical issues involved in the use of animal biotechnology: Is it acceptable to create an animal that feels no pain through the use of animal biotechnology? Why or why not?

What are the benefits and to whom?

What about the concept of "telos," or an animal's true nature? Making an animal that is incapable of feeling pain goes against an animal's telos. Is this acceptable?

An animal has been used to create a promising new drug for the treatment of many types of cancer.

However, the process involved might bring pain to the animal used to make the drug. Is this acceptable considering the end result? Why or why not?

Should we make the animal incapable of pain in this case? Why or why not?

Companies have begun to apply for and receive patents to protect their often expensive investments in animal biotechnology. The average investment these companies make is about $55 million over seven to nine years. Based on this expense, should companies be able to patent a living organism? Why or why not?

How is this different from a patent issued to a company for a nonliving product, such as new software?

What alternatives can you think of for the patenting system?

Chapter 5

The Impact of Livestock Development on Environmental Change

Environmental change, over the next 50–100 years, due to the warming effect of the accumulation of gases in the atmosphere will clearly influence man and his well-being and will introduce change to many areas of the world. Although there has been modelling of future climate change, the major observation is that these models are as yet producing limited useful information. Computers have the ability to consider a multitude of variables but the computer is limited by the knowledge base and magnitude of the unknowns.

In fact, it can be quickly ascertained that modelling, at best, is defining what is not known in terms of what will effect change in temperature, rainfall and sea levels, let alone sea currents, wind, sunshine hours, soil moisture and the incidence of pests and diseases of man, animals and plants. Some prediction (no matter how uncertain) are welcomed by the population at large (e.g. increased environmental temperatures in Europe) but the uncertainties indicate that for any benefits there will also be major disadvantages, both economic and social.

For example, high temperature and increased evaporation rates and the lower resultant soil moisture content, may result in the death of large areas of planted and natural forest in Northern Europe; rain in previously dry areas may lead to large scale soil erosion and wide scale flooding of major agricultural lands in the fertile delta country will occur if sea levels increase by 0.3 metres by 2050. The major conclusion is that disadvantages are likely to outweigh any advantages and the unknowns make it essential to put into action technologies to slow green-house gas emissions and stabilise atmospheric gases as soon as possible. The long lag time between gas production, mixing in the atmosphere and therefore warming together with the reluctance of governments to put into practice legislation to limit, in particular, carbon dioxide production from fossil fuels, suggests that there will be a rise in world environmental temperatures of 0.5 to 1°C in the next 25–50 years. A major point is that no one country can point to a low level of "environmentally dangerous gases" production as being a reason for non-compliance with a general lowering of emissions. Virtually all countries contribute a small proportion of the greenhouse gas and it is, therefore, necessary for all countries to take action. In other words it needs action world wide and "every little bit helps".

A distinction must be made between temperature rise due to gas accumulation in the atmosphere and depletion of the ozone layer through reaction with atmosphere contaminants. The two are linked but the depletion of the ozone layer (this layer protects the animals/plants from deleterious dose rates of UV irradiation) is largely a result of reaction of the ozone with atmosphere contaminants. Ozone depletion is not highly related to the subject of environmental change and gas accumulation in the atmosphere as discussed here. However, increased ultra-violet radiation at the earths surface will have major detrimental effects on plant growth.

The Greenhouse Effect

A Simple Description

The greenhouse effect, or increasing world temperature, is clearly ascribable to the major industrial countries as some 50% of the increased retention of energy by the atmosphere is a result of the accumulation of carbon dioxide from combustion of fossil fuel. Industrialised countries presently use 70% of the world's oil production and it has been much higher in the past. The other gases that contribute to increasing temperatures arise from a variety of activities and some of the gases have been created by man. Methane is an important component of the increasing gases in the atmosphere and is the one most associated with animal agriculture. Methane has a thermogenic effect some 4–6 times that of carbon dioxide.

The rates of accumulation of methane and carbon dioxide in the world's atmosphere have changed dramatically in the last 10 years. Prior to this, the rise in world temperatures and composition of the atmosphere had changed little, but is now in what appears to be an exponential period. Undoubtedly the contamination of the atmosphere with carbon dioxide, methane and other gases must be stabilised or the future of the earth is threatened.

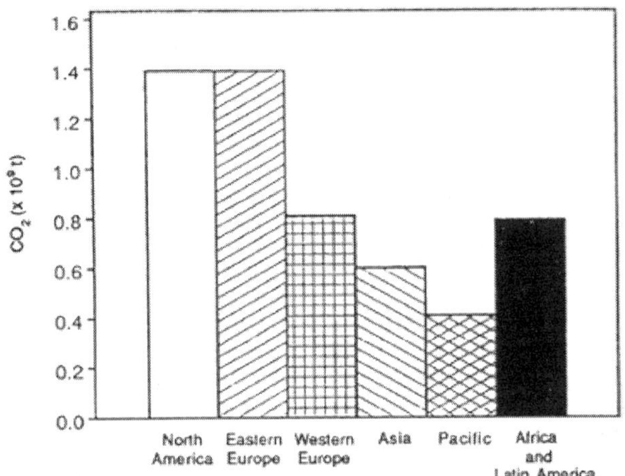

Figure: *Relative contribution by continent to the emissions of carbon dioxide (Source: World Resources Institute).*

Methane production appears to be a major issue although it presently contributes only 18% of the overall warming. It is accumulating at a fast rate, and is apparently responsible for a small proportion of the depletion of the protective ozone layer. Methane arises largely from natural anaerobic ecosystems, rice paddies and fermentative digestion in ruminant animals.

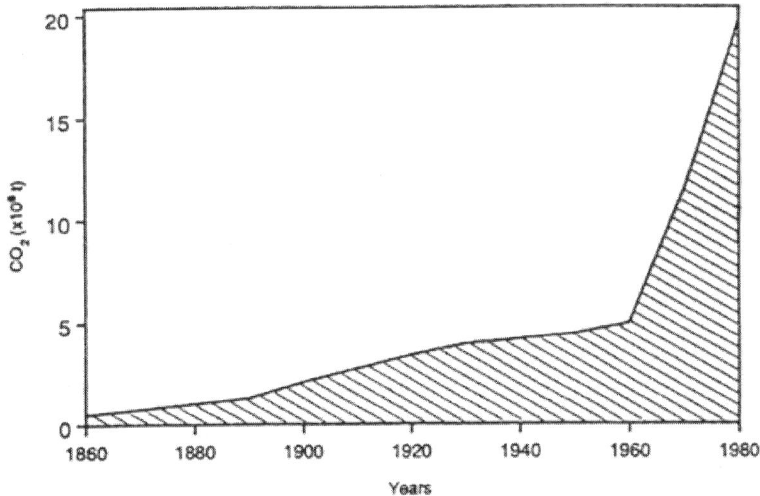

Figure: *Trends in emissions of CO_2 (Source: World Resources Institute).*

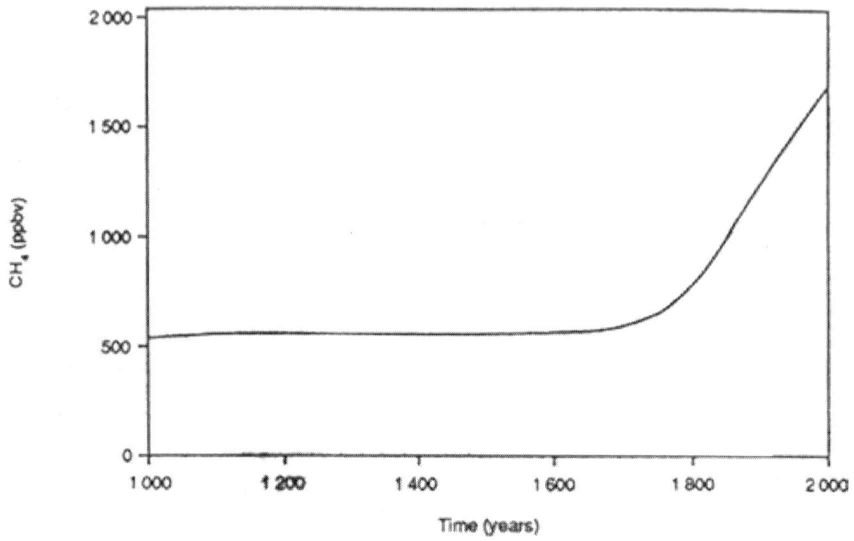

Figure: *Trends in atmospheric methane accumulation (Khalil and Rasmussen, 1986).*

Animal Agriculture and the Greenhouse Effect

Animal production plays four important roles in the release of gases into the atmosphere:

- directly through production of methane in fermentative digestion of ruminants,
- indirectly when a proportion of the faecal materials decompose anaerobically,
- indirectly through CO_2 production from fossil fuels to provide the production and marketing infrastructure and inputs such as motorised transport, fertilizers, herbicides and insecticides,
- through the clearing of forests and range lands, the timber on which was a natural sink for carbon dioxide.

Ruminants in 'natural' production systems are inefficient and, in general, production increases depend on an expansion of numbers. There is a growing appreciation that efficiency per animal can be improved many fold with simple technology inputs which would have an impact on all four aspects of the contributions to global warming discussed above.

The most important approach to be discussed in relation to the amelioration of greenhouse gas production by ruminants is to increase the efficiency of animal production from available resources and develop the capacity to produce more from less animals.

Requirements For Animal Products

Developed Countries

With largely stabilised human populations in the industrialised countries and a high standard of living generally, the demand for animal products has plateaued or declined and the emphasis in animal agriculture is on the production of higher quality products. This has led to a dependence of ruminant systems based on concentrate feeds, a higher production rate per animal and reduced animal numbers.

For example, there has been a marked increase in milk production through "better nutrition" relying on concentrate diets and genetic improvement of dairy cows. The result has been a marked reduction in the total number of animals required to supply local milk requirements. These increases in production per animal in North America and EEC countries that have arisen from simple technology inputs are indicated.

The demand for food in the industrialised world is stable but production is increasing at 1.5% per annum (FAO, 1986). Policies to take land out of farming or to limit farm management options and animal numbers has reduced overall animal numbers, but meat/milk production has been maintained by technologies that increase the efficiency of production.

From a global warming view point this is mostly advantageous, although there is a very high environmental cost of feeding high grain based diets to ruminants. Such concentrates can only be produced by high fossil fuel inputs and is often only economic because of subsidies. Les intensive production systems depending on grass or grass products in developed countries exhibit some of the same problems of low efficiency as those animals in developing countries which are restricted to roughage based diets.

The Developing Countries

Throughout the last 30 years crop and livestock production has more than doubled throughout the third world, although there are large differences between regions, however, increases in human population, Urbanisation and improved income levels have increased demand for food to such an extent that surpluses are still rare. Imports of meat, milk and cereal grains in most developing countries are increasing by about 10% per annum. The demand for animal products will continue to increase for some time, if the patterns of food consumption by people in developing countries follows the patterns that occurred in developed countries.

Increases in animal production in the developing countries has been mainly a result of increasing animal numbers (Jackson, 1981). The lack of increase in efficiency of animal production is well documented and is emphasised by the average milk yields per cow over the 10 years from 1976–1986 (Brumby, 1989). This low productivity is exacerbated by long calving intervals and a late age at puberty in the cows in developing countries. It should be emphasised however, that it is also a feature of ruminants fed low quality forages in any country.

New feeding strategies for animals fed on low quality forages (e.g. crop residues, tropical pastures etc.) coupled with better genotypes, improved management and disease control, particularly in India (NDDB, 1989), has changed this situation enormously.

Table: *The change in the average milk yield per cow in industrialised and third world countries.*

Country/Region	***Average yield/ cow(kg/year)***		***Percentage increase (%)***
	1976	***1986***	***(1976 to 1986)***
North America	3,250	5,200	60
EEC Countries	2,900	4,100	35
Asia	620	700	13
Africa	322	354	7

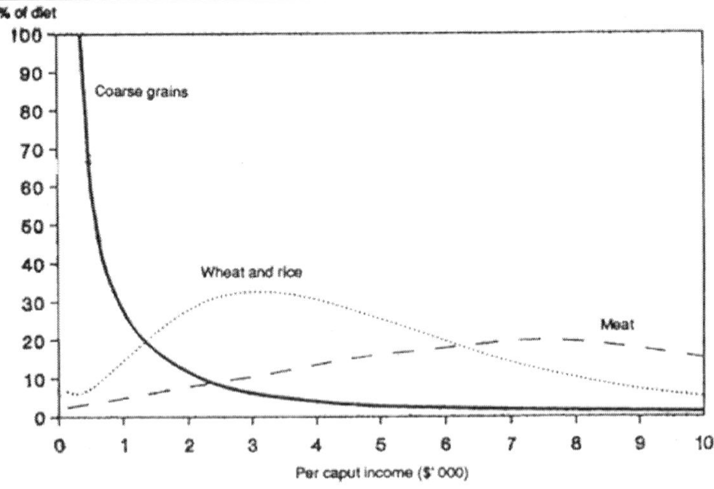

Figure: *Food consumption/percentage of diet for meat, wheat & rice and coarse grains. (Marks & Yetley, 1987 and Brumby, 1990).*

A major difference in approach to feeding ruminants in tropical developing countries that has been recently developed has the potential to revolutionise ruminant production from forages. For example enormous increases in milk production can be achieved in the tropics without the use of 'fossil-fuel-expensive' grain based concentrates, relying rather on byproducts of agriculture.

These are the only truly available feed resource for the large ruminant populations in the foreseeable future. The low inputs of concentrates into such systems and levels of production that rival those achieved in the industrialised world, provide major indications for change in management of both developed and developing countries alike.

Population Densities of Large Ruminants

The above discussions suggest that ruminant populations are not likely to increase in the industrialised world and, with ever increasing technology inputs, numbers are set to decline (as has already occurred with milking herds) - but continuing technology inputs will be highly fossil fuel dependent.

Conversely, there is likely to be a huge increase in demand for food products in the countries that are developing, of which a greater proportion of the demand is likely to be for animal products. To meet this demand local production must be increased to an extraordinary extent and it is most desirable that the impact on the environmental contamination is minimised.

The ranges of Y_{ATP} are shown for: The demand for draught power in countries with large numbers of small farmers is likely to expand rather than contract in the future and must not be left out of any considerations. This large group (80 million draught oxen in India alone) receives the least inputs and yet they are probably one of the major factors in food production and they limit the use of costly fossil fuels in developing countries.

Environment-friendly development of livestock production systems demand that the increased production must be met by increased efficiency and production per ruminant and not through increased numbers. The need to increase numbers would put huge pressures on many resources including forests and land that might be afforested.

Methane Production

Methane Production from Ruminants

World ruminant population densities and estimated methane production rates are shown in table in comparison to some monogastric species including man.

Global Methane Production

Methane gas accumulation is at a rate of 1% per annum and methane contributes 18% to global warming. Ruminants from all species produce a relatively small proportion of the global production i.e. 15–20% of the total methane generated.

However, the domestic ruminants represent one of the few sources that could be manipulated. It is estimated that beef and draught animals contribute 50%, dairy cows 19% and only 9% is from sheep (Crutzen *et al.*, 1986).

The methane is generated largely in the fermentative digestion of feed by microbes in the rumen. The data in table indicate an approximate even split of numbers of ruminant animals between the developing and developed countries; that large ruminants contribute the greatest to world methane accumulation, and that other domestic and wild herbivores and man contribute an insignificant proportion.

The simple approach to the calculations adopted by Crutzen *et al.* (1986) may, however, be a little misleading although they are the best estimates available. It is most important to emphasise that it is the rates of methane production per unit of product produced over a life time which is important in identifying where it is possible to make a major reductions in methane emissions. This is the major philosophy developed in the rest of this presentation.

Productivity of Ruminants Fed Poor Quality Forages

The vast majority of ruminants in developing countries and a major proportion of the national herds of industrialised countries are supported on the by-products of agriculture or graze forages of relatively poor nutritional value.

In general, growth rates, milk production and reproductive rates in these systems depending on forage of variable quality are extremely low compared with the genetic potential of these animals (mostly about 10% and rarely exceeding 30%). Mostly cattle grow to maturity or slaughter weight over 4–5 years, cows produce their first calf at 4–5 years and then, on average, every two years.

Milk production on these feeding systems is often below 1000 litres/ lactation. Cows may be kept largely to produce draught oxen and in some specialised systems they are kept for the production of dung (which is valued as a fuel) and a number of other minor purposes (e.g. as a investment, for recreation and for religious purposes).

Table: *Estimates of methane emissions from animals (adapted from Crutzen* et al., *1986).*

Animal Type and Region (X10^{-6})	***World Pop.***	***CH_4 Prod. (kg/hd/yr)***	***Total CH_4 Prod.***
Cattle:			
Developed countries	573	55	31.8
Developing countries*	653	35	22.8
Buffaloes	142	50	6.2
Sheep:			
Developed countries	400	8	3.2
Developing + Australia	738	5	2.4
Goats	476	5	2.4
Camels	17	58	1.0
Pigs:			
Developed countries	329	1.5	0.5
Developing countries	445	1.0	0.4
Horses	64	18	1.2
Mules & Asses	54	10	0.5
Humans	4670	0.05	0.3
Wild ruminants and large ruminants	100–500	1–50	2–6
Total			76–80

* includes Brazil and Argentina

** total estimate for emissions from domestic animals has an uncertainty factor of ± 15%

Slow growth, low milk yield and poor reproductive performance results in poor feed conversion and a large methane output relative to product output.

Methane Production in Ruminants Fed 'Poor Quality' Forages

Methane output relative to product output of ruminants depends on two factors:

- the efficiency of fermentative degestion in the rumen, and
- the efficiency of conversion of feed to product (e.g. milk, beef, draught power).

Efficiency of Rumen Fermentation

Digestion of feed by ruminants depends on a diverse group of microorganisms in the rumen. These organisms ferment feed materials into volatile fatty acids (VFA) a process that produces methane (CH_4), carbon dioxide (CO_2) and utilises the energy (ATP) derived to convert feed to microbial cells. The partitioning of the feed components into VFA or microbial cells and the release of CH_4 and CO_2 depends on a number of factors. Feeds that allow a high efficiency of microbial cell synthesis produce low amounts of methane per unit of feed digested.

With cattle fed on a poor quality forage a number of essential microbial nutrients are usually deficient in the diet and microbial growth efficiency in the rumen is low. In these conditions CH_4 produced may represent 15–18% of the digestible energy and correction of these deficiencies may reduce this to as low as 7%. The relationships between products of fermentative digestion and the efficiency of the microbial ecosystem in the rumen.

Efficiency of Feed Utilisation by Ruminants Fed Crop Residues or Other Fibrous Feeds

Liveweight gain. Research in the past 20 years has clearly illustrated that supplementation of cattle on low quality forage based diets effects productivity through increasing efficiency of feed utilisation. A mixture of nutrients as can be supplied for instance in a molasses urea multi-nutrient block/lick ensures an efficient microbial digestion in the rumen. Also small amounts of protein meal that are directly available to the animal (i.e. bypass protein) stimulate both productivity and efficiency of feed utilisation. Traditional feeding standards are based on the metabolisable energy (ME) content of a feed. The general relationship between ME/kg of feed and growth (g gain/unit of ME intake) are shown. The results of a number of feeding trials with cattle on straw or low quality pasture and silage based diets supplemented with protein meals are shown in the same.

The efficiency of growth and methane production is shown. These data clearly show the massive reduction in methane production per unit of liveweight gain that is possible by using a strategic supplementary feeding system that accommodates the requirements of the rumen organisms and balances the absorbed nutrients to the animals requirements. The data show the effects on methane production of balancing the rumen and for protein supplementation calculated for experimental data with growing animals of Saadullah (1984). This indicates the massive potential reduction in methane production per unit of liveweight gain that can result from supplementation (Leng, 1989). Provision of molasses urea blocks to draught oxen would have a major effect on methane production, reducing it to half the present production rate.

Milk Production

The same principles of supplementary feeding have been found to stimulate milk production of dairy animals. Without going into detail, the methane produced per unit of milk produced under traditional feeding of local dairy cows or imported Friesians in India or Friesians under temperate country management production in relation to lifetime milk production and includes the effects of supplementation on age at first calving, intercalving interval and improved milk yield of supplemented animals. The relationships found in practice with cattle fed on straw or ammoniated straw with increasing level of supplementation. Australia (Perdok *et al.*, 1988), Thailand (") (Wanapat *et al.*, 1986) and Bangladesh (¡%) (Saadullah, 1984).

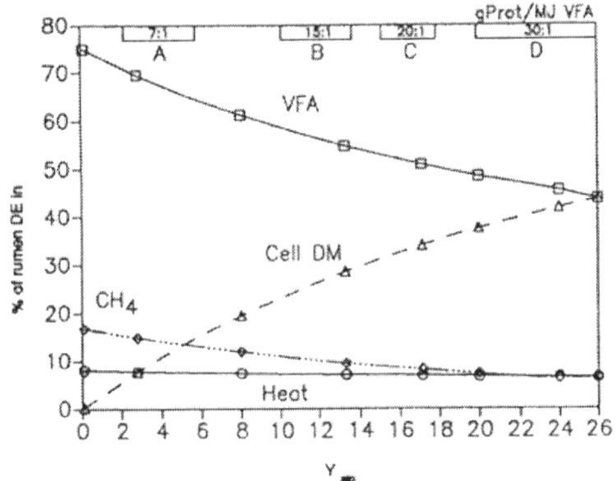

Figure: *Schematic relationship between diet quality (metabolisable energy/kg dry matter) and food conversion efficiency (g liveweight gain/MJ ME).*

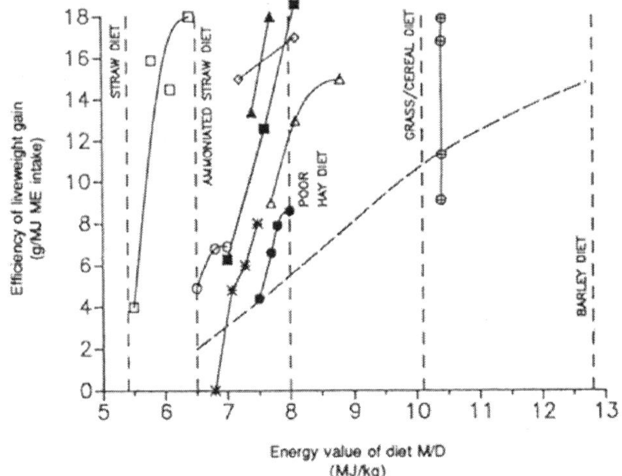

Figure : *Schematic relationship between diet quality (metabolisable energy/kg dry matter) and food conversion efficiency (g liveweight gain/MJ ME)*

Recent relationships developed for cattle fed silages supplemented with fish proteins (Olafsson and Gudmundsson, 1990) (Å) and tropical pastures supplemented with cottonseed meal (Godoy and Chicco, 1990) (*) are also shown. This illustrates the marked differences that result when supplements high in protein are given to cattle on diets of low ME/kg DM. The relationship shown by a broken line is based on the metabolisable energy system in practice in the U.K. (Webster, 1989). The other relationships are results from results quoted. Perdok *et al.*, 1988, Saadullah, 1984, Wanapat *et al.*, 1986, Godoy and Chicco, 1990 (²%) and Olafsson and Gudmundsson, 1990 (f&).

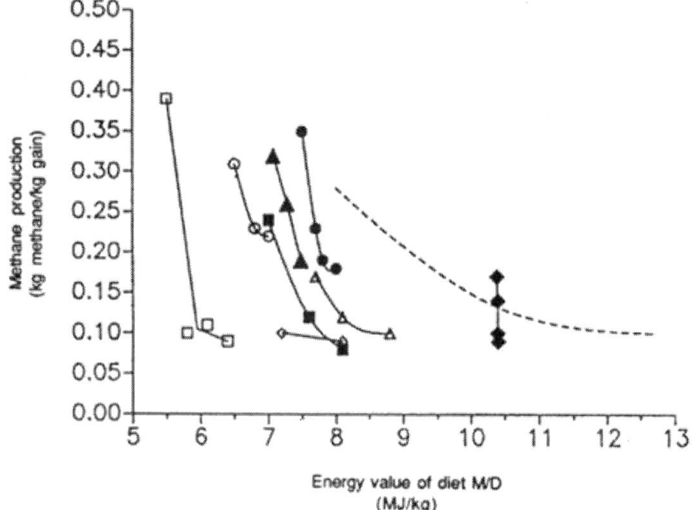

***Figure:** The relationship between the metabolisable energy content of a feed (M/D, MJ/kg) and the methane produced/kg gain.*

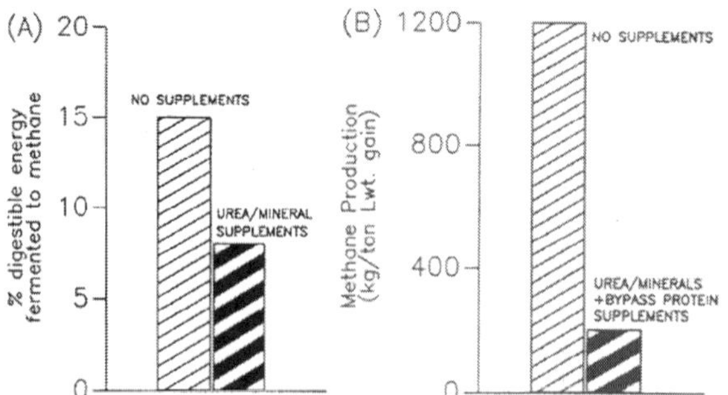

***Figure:** (A) The effects of improving the efficiency of rumen fermentative activity on methane production/kg of digestible energy consumed. (B) The production of methane/kg gain in supplemented cattle (feed conversion efficiency (FCR) 9:1) or unsupplemented cattle (FCR=40:1) fed straw based diets (after Saadullah, 1984).*

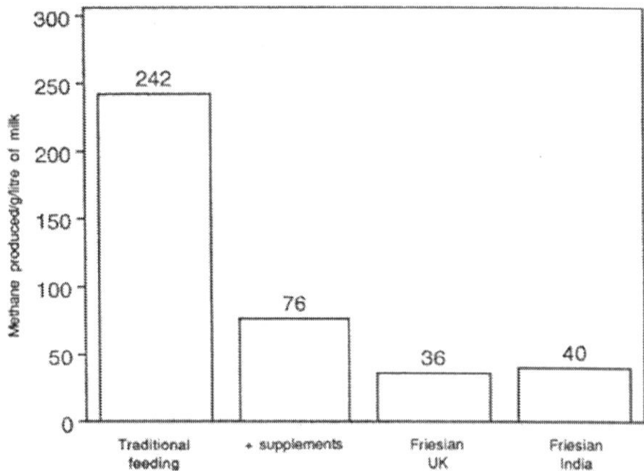

Figure: *The methane produced per unit of milk production in unsupplemented (fed traditionally) or supplemented (new feeding systems) cows in India with moderate levels of production.*

Methane gas content of the atmosphere is increasing at 1% per annum. To stabilise methane in the atmosphere, global methane production needs to be reduced by 10–20%. Large ruminants produce some 15–20% of the global production of methane. Ruminants on low quality feeds possibly produce over 75% of the methane from the world's population of ruminants. Supplementation to improve digestive efficiency in these animals could at times halve this methane production per unit of feed consumed. Together with supplementation to improve efficiency of feed utilisation and increase product output may thus reduce methane production per unit of milk or meat by a factor of 4–6.

Provided animal numbers in national herds are decreased as demand is met, the production of methane from the large populations of animals fed poor quality forages could be reduced to below 50% and perhaps even to as low as 20% of its present rate. It is probably not feasible to restructure the cattle industries of the countries involved.

The socioeconomic implications are difficult to predict, but the start made in India with up to 1 million head of cattle, and with a potential to reach 7–10 million animals in the immediate future, is highly desirable because it increases milk production with a concomitant decrease in actual methane production and methane production per litre of milk. Social, anthropological, economic and political considerations have been the major determinants of 'aid programmes' but the growing 'environmental crisis' due to increasing content of gases in the atmosphere leading to global warming is likely to dominate many issues confronting aid-agencies in the future. The achievements in India can be repeated in most countries that depend on poor quality forages and even silages for ruminant production. India is fortunate in having large amounts of protein meals as by-products which can be used for animal production. In other countries protein meals may be scarce or under-utilised. The future for increased animal production in these countries is to identify protein sources and proceed to find economic mechanisms for producing meals that contain a high proportion of bypass protein. The developed world may need to encourage this by subsidising such activities or by refraining from draining _these protein sources from the developing world to support subsidised over-production in the industrialised countries. The world trade in protein meals is shown in table.

Table: *The trade in oilseed cakes between industrialised and third world countries (Borgstrom, 1980).*

	Imports	*Exports*	*Net imports*	*Net exports*
Industrialised countries	18.7	9.4	9.3	-
Developing countries	1.7	10.8	-	9.1

The increases in efficiency of animal production that result from the new nutritional strategies will, if widely applied, increase animal productivity to the extent probably needed by the expanding human population in developing countries. However, it needs a retraining of technologists in these countries to absorb the new concepts and a move away from temperate country teaching of ruminant nutrition. The tropical climate can be used to the advantage of the developing countries where ruminant production can be more efficient.

Acceptance of these feeding strategies could reduce the need for land clearing and pasture establishment in the fragile areas of the world which have been so prone to erosion following clearing. It may also allow for considerable change in the use of such pastures with reforestation as a highly desirable first option. In addition, reduced world ruminant populations would reduce incidental release of methane from decomposing dung and reduce the inputs of fossil fuels required in much of the infrastructure.

The Conservation of Animal Genetic Resources in the Developing Countries: a Practical Way Forward

The Economic Importance for Genetic Resources

Important economic gains may accrue from the appropriate choice of the livestock genetic resources correctly utilised in a given production system. This is illustrated by results a trial on dairy cattle crossbreeding strategies undertaken in Brazil, in which the accumulated lifetime performance records of 527 females have been recorded over the past 14 years on 67 farms. Six levels of crossbreeding between Holstein-Friesian (HF) and zebu Guzera (Z) were included with the HF genetic composition ranging from 1/4 to e" 31/32 HF fractions. A summary for the first 9 years.

Farms were grouped into two management classes for analytical purposes. The F1's were the most profitable group under both management regimes and for this reason was taken as a reference point to express relative performance of the other crossbreeding strategies. Rotational crossing of HF sires for two generations followed by one generation of zebu sires (HF-HF-Z) was the second best alternative in the high management group. Upgrading to HF was as profitable as the HF-HF-Z rotation under high management, except under the higher fat and protein pricing which would be fairer to farmers. However in the low management grouping, upgrading to HF would have had disastrous consequences due to low production, high mortality and low culling rates. A new breed developed from *inter se* matings would require a high selection intensity to reach the same profitability as the F1. Choice of strategy would, of course, depend not only on performance but also on the actual possibilities of implementing each crossbreeding scheme (Madalena, 1989).

The above example shows that by using the right type of animal, without changes in either nutrition, health or other inputs, profit was considerably increased. Conversely, losses resulted from the inappropriate choice of crossbred. This, of course, does not mean that environmental factors should not be improved, but rather that both the genetic and environmental components should be considered in unison. The choice of germplasm is an integral element in the production system and it must be carefully matched to the other available inputs. It might be appropriated to discuss the implication of the genotype x environmental interactions that are seen in the Brazilian trial. The view is sometimes expressed that there is no need to worry about genetics until the management is sufficiently improved to allow full expression of the existing available genetic potential.

This view, however, fails to recognise that some genotypes have higher potential in a favourable environment but a lower potential in more stressful conditions. Therefore the notion that there is a genetic potential for each level of management is conceptually and practically more accurate (Falconer, 1960). As illustrated by the results shown ignoring genetic differences would be an unwise decision in any development programme.

Table: *Profit per cow per day of herd-life under alternative strategies of crossbreeding of Holstein-Friesian (HF) x zebu (Z).*

	Management Level*							
	"High"				"Low"			
Price-cost**	A	B	C	D	A	B	C	D
F1 profit***	1.8	2.8	3.5	2.5	4.6	5.5	5.1	5.4
Strategy	percent of F1 performance							
HF-HF-Z rotation	75	72	84	81	48	51	52	47
HF-Z rotation	41	53	63	57	59	60	61	63
Upgrading to HF	75	57	84	80	-21	-12	-10	-28
New breed (5/8HF:3/8Z)	-18	1	23	14	30	33	34	55

* Characteristics of management levels:	"High"	"Low"
Mean first lactation milk yield (kg)	2450	1731
Mean first lactation length (days)	283	309
Mean first calving interval (days)	402	561
Concentrates fed, kg/cow/day	4.5	1.6
All milked 2x/day with calf suckling to stimulate let-down.		

All milked 2x/day with calf suckling to stimulate let-down. ** A: 1980 to 1985 prices (protein not paid for) B: Fat differential tripled and protein paid at same rate. C: Cost of concentrates halved. D: Beef value of animals doubled. *** Profit expressed in kg milk/day. Price of 1 kg milk (3.3% fat) = US$ 0.16.

Source: Madalena et al. (1989).

Genetic Variation a Natural Resource

Historically, man has reshuffled genes from different livestock populations by crossing, selection and inbreeding. Biotechnology now offers powerful new methods to change the genetic composition of animals. However, because genetic material cannot be synthesized, improvements will still be restricted to the obtaining of the best possible combinations of existing DNA. Therefore, animal genetic resources constitute an indispensable natural resource must be properly managed for efficient production now and to be preserved for future use. Although germplasm introduction has traditionally been a common practice in animal production, it was only really in the last two decades that genetic variation came to be viewed as a natural resource (Dickerson, 1969). Commercial use of genetic variation between populations has increased dramatically in some species as has research, primarily in developed countries, regarding the its theoretical aspects and the evaluation of alternative breeding plans.

Although migration also plays an important role in the genetics of animal populations in developing countries, research generally did not proceed, or even accompany, commercial trends. Genetic resources in these countries have not been adequately evaluated or fully utilised and in some cases are threatened with extinction without being properly described. The following activities might be listed for action on genetic resources management:

- description
- evaluation
- utilisation
- conservation.

FAO's programme has recently been described by Hodges (1990) and I will briefly refer to the above in the following sections, concentrating on issues rather than methodology.

Description

A brief description of the world's main livestock species and breeds is given in Mason's (1988) "World Dictionary of Livestock Breeds". A Global Data Bank of Animal Genetic resources has been established jointly by the European Association of Animal Production and FAO at the Institute of Animal Breeding and Genetics, Hanover School of Veterinary Medicine (Simon, 1990). Information from the bank is publicly available. A questionnaire was developed to collect information on breed origin, numbers, phenotypic description, uses, management systems and relative performance compared to standard breeds. The work was initiated with cattle, buffalo, sheep, goats, pigs and horses, however, questionnaires for poultry and Andean cameloids are in preparation.

The databank currently has information on 658 breeds from 25 countries (mostly European, USSR and China) and procurement of data from other countries is being sought. In many cases such information is locally available although it is not generally readily accessible at hand, so compiling it in one databank requires motivation and effort, especially for completing the questionnaires. In other cases, special field surveys may be required to obtain the relevant information.

Estimates of genetic distance between breeds would be valuable in formulating conservation policies, thereby allowing a more rational selection of breeds requiring preservation. Where possible DNA description is preferable to indirect description involving gene expression (Primard, 1985). Trinity College, Dublin, is undertaking a study to measure genetic distances between 6 european and 8 tropical breeds (from India and Africa) of cattle. DNA finger-printing based on endonuclease restriction length fragments is being evaluated and both mitochondrial and nuclear DNA concentrates are being extracted locally and transferred to Dublin for analysis (D. McHugh and R. Loftus, personal communication).

Evaluation

The systematic introduction and evaluation of germplasm is a common practice in plant breeding but less so in animal breeding, except perhaps in poultry. However, important work has been undertaken such as the trials established by the Meat Animal Research Centre in Lincoln, Nebraska, where a large number of cattle breeds have been evaluated over the past 20 years; the early Argentinean beef breeds comparison trial in the 1960's; the evaluation of European breeds in Great Britain and the FAO trial in Poland to compare Holstein-Friesian strains from 10 different regions.

The evaluation of germplasm is just one particular aspect of the overall research requirements. One hesitates to emphasize germplasm evaluation since the benefits would appear to be obvious. Yet, there have been many cases where recommendations based on opinion, rather than of results, have had disastrous effects. This is particularly the case where livestock importations have been involved and Vaccaro (1990) indicated that european dairy breeds are not sustainable. Also the general belief that 5/8 European x 3/8 zebu cattle crosses were the best combination for tropical environments has caused unnecessary delays in new breed developments (Madalena, 1989) and, in spite of early warnings (eg. McDowell, 1972), it has only been possible to put the matter in its proper prospective in the light of recent experimental results.

Another example of the importance of germplasm evaluation may be found in the Criollo, cattle that were originally introduced into Latin America, and that have adapted over the centuries. The Criollo have now almost completely subsumed by the zebu in the lowland tropical areas, a process that intensified in the 1920's (de Alba, 1987). Recent research has shown that purebred zebus are no more productive than the purebred Criollos, although the F1 crosses are superior to both, due largely to heterosis expressed in reproductive efficiency and other economically important traits (Plasse, 1989).

Therefore, a crossbreeding policy would have been indicated instead of the breed substitution. As de Alba (1987) pointed out, farmers and ranchers impressed by the crossbred's performance attributed it to the new breed and not to heterosis - which is a more difficult phenomenon to grasp. The result was that a population of over 100 million head were pushed in the wrong genetic direction. Had the research been done 70 years ago and the results explained to farmers, this situation might have been avoided.

To be useful, germplasm evaluation must be comparative. The breeding alternatives should be compared over the same environments using appropriate designs, which include sufficient numbers of representative animals and sires. Lifetime performance needs to be recorded, since both survival and herdlife are major components of the overall economic performance of breeds in stressful environments, and large differences in these traits are expected for diverging genotypes. With ruminants, this means that evaluations have to be in the order of 10 years. Shortcuts are not available and would only lead to underestimation of the real economic differences between alternatives.

Germplasm evaluation should, wherever possible, be conducted under commercial management rather than at experimental stations. In developing countries government farms are often inadequately funded, poorly administered and it may not be easy to simulate the variety of management and socioeconomic situations found in the private sector. For example, it is unlikely that administrators would allow the poor husbandry conditions found in some commercial farms. While the basic components of economic performance should be recorded on farms, some investigation of the more sophisticated traits may need to be undertaken to understand why the observed differences occur. For example, product quality may be conveniently analysed at a central laboratory, although getting the samples to it may be not an easy task.

Operationally, on-farm trials are usually cheaper, since farmers do not carry bureaucratic overheads. It is also less likely that administrative changes would disrupt long-term experiments in farms as can happen in experimental stations. On-farm recording, however, although not expensive, requires organisational and logistical support along with careful supervision.

I submit that the real opportunity cost of on-farm germplasm evaluation is not high. The costs of producing the necessary animals, distribution and performance recording may be largely recovered from production through some form of share-farming schemes as was done in the Brazilian trial. Staff and facilities for supervision, analysis and to interpret results would be available at research institutes/ universities in many countries, although they are likely to be overburdened with other activities. Therefore some support would be needed to strengthen laboratories, hire technical staff and provide minor, unsophisticated equipment and reagents.

Conservation

Characteristics of both species and breeds result from the expression of not only individual genes but also from gene combinations which interact to influence the physiological processes. Genes and gene combinations are renewed by reproduction, however, they may also be lost through either the extinction of the population or replacement by other genes or gene combinations. Selection and crossing will cause genetic change in a given direction, conversely, a low population size and inbreeding may result in the random loss of genetic material. Depleting genetic variation will restrict the choice of available genetic material for use by future generations. Should new circumstances arise, requiring animals with new characteristics then the genetic material would not be available to develop it. Conservation of animal genetic resources must be seen as one aspect of the wider problem of maintaining bio-diversity. Present increased resource utilisation conflicts with possible future needs.

As the decision making process becomes increasingly centralised through the development of rural organisations, communications, education and propaganda; and the increasing use of reproduction tools like artificial insemination and embryo transfer, so the power of man to change the genetic make-up of livestock will increase dramatically.

Cases are already known of breed replacement on almost a continental scale within a few decades. Some preservation is clearly needed to allow future generations the opportunity to be able to use the genetic resources that are currently available today. Thus, conservation of genetic resources could be considered as an insurance. Studies have shown that the economic return is very high when the preserved germplasm is extensively used (Smith, 1984). However, the likelihood of the preserved resources being useful in the future is, by definition, not known so the decision on how much to spend remains largely a subjective one.

At present, there are two basic ways of preserving genetic material for further utilisation, either as live animals or as deep frozen semen or embryos. Freezing of ruminant semen has been used for a long time and embryos for the past decade. Freezing of pig and horse semen is a more recent development, although freezing pig embryos is not yet possible. Freezing of poultry semen is not yet commercially feasible. The conservation, or reactivation, of other cells, chromosomes or DNA should still be considered at an experimental stage (Brem and Brenig, 1990).

Because biotechnology is moving so fast, techniques may rapidly evolve, so it may be appropriate to set a conservation planning horizon of, say, 25 years. Future users would then decide if and how to reactivate stored material; continue to preserve it (or indeed to throw it away!) in the light of available techniques at the time. A short planning horizon is not reasonable, because most of the conservation cost is incurred initially, to establish herds or to freeze the semen/embryos.

Judgement is required to compromise between the size of the genetic sample, the number of unit to be preserved and the cost of preservation cost. Based on inbreeding considerations Smith (1984) suggested storing 100 semen doses each from 25 unrelated sires and 25 embryos from each of 25 unrelated females donors and sires per breed. Indicative costs of semen preservation for 25 years would be US$126,000 per breed and US$158,000 for embryo preservation. Sixty-five and 86 percent of these costs, respectively, correspond to the initial collection cost while maintenance costs are relatively inexpensive. Although semen is cheaper to preserve, reactivation would be faster if embryos were also available.

Alternatively, a herd/flock of 25 breeding females could be kept with no inbreeding over a 25 year period, provided controlled matings or artificial insemination was possible. Under pasture management this should not be expensive. Thus, *in situ*, conservation appears to be simpler than cryo-preservation. However, the risks of losing genetic material due to constraints in population size are higher for live animals. Disease outbreaks, droughts, broken fences, periods of poor management and the intrusion of unwanted sires in the herd might not be rare occurrences and would need to be guarded against.

Therefore, each conservation method has its merits and weaknesses which need to be assessed for each particular situation. For example, some countries have established embryo manipulation facilities and these could be made available to other countries in the region, such as, the proposal for Regional Gene Banks promoted by FAO.

In general, conservation of genetic resources is a long-term activity but there may be special cases in which its objectives may complement shorter-term development goals i.e. in the case of breeds that are commercially under-utilised, in spite of their potential to improve the economic efficiency of present production systems. Yet these valuable breeds may not be preferred by farmers or decision makers for a number of reasons, including: lack of research regarding their potential contribution, propaganda and/or vested interest. Germplasm from these breeds should be evaluated in the context of designing breeding strategies, including breed development, which would lead to their commercial conservation.

Development and conservation must go together. In fact, some contend that under-development leads to loss of natural resources (D. Wood, personal communication). There is also, however, a moral issue: it is difficult to think of saving endangered breeds when confronted with children suffering from hunger caused by famine. Yet, we do not wish to hand over to the next generation a world depleted of the majority of it's natural resources. Essentially mankind has the necessary genetic resources in terms of breeds and individuals to meet the challenge of feeding itself in the future (Cunningham, 1991). Some conservation effort is justified to sustain development in the long term and a balance will have to be established. It is certainly not logical to preserve resources for future use when we do not properly use them now. This would be like spending all your money insuring a car and then not be able to buy the petrol to run it.

Chapter 6

Field Development of Livestock Projects

The livestock industries of Third World countries frequently underpin their agricultural production systems. Livestock production is often the main source of disposable income and realisable capital assets amongst smallholder and pastoralist communities and provides essential non-cash benefits including draught power and manure. Dairying in particular offers a reliable and regular source of income, especially for poorer rural households, disadvantaged social classes and women. Despite this relative importance, investments in large ruminant development in both smallholder, livestock/crop farming systems and in pastoralist based systems have been perceived as only marginally satisfactory. The results of the World Bank lending programme for livestock development highlight this point. After a peak of US$340 million per year over the period 1974–79, World Bank lending for livestock development declined to US$240 million per year over the period 1980–85 and is now about US$100 million per year . The reduction in funding for rangeland based livestock production was even sharper than these figures would indicate with a corresponding rise in funds targeted at the smallholder livestock sector. Within this lending programme, cattle development accounted for two-thirds of funds committed between 1959 and 1983 and in 1985 represented about 50% of the livestock loan portfolio.

The perceived overall failure of livestock projects, and by implication, of large ruminant development, is not supported by results. The 1988 World Bank annual review of project evaluation results showed the following outcome for the period 1974–1988.

Had this analysis taken into account the non-cash benefits from livestock production, which often account for 50% of output, livestock projects may well have been assessed much more favourably. In fact, the 1985 review of smallholder livestock projects concluded that the performance of individual livestock projects in the World Bank portfolio was generally satisfactory in most regions of the world, excluding West Africa and East and Southern Africa. The conspicuous failure of several pre-eminent African livestock development projects in the last two decades has had a disproportionately negative impact on livestock development in general.

Running counter to this experience is the growing evidence of linkages between improved livestock production and enhanced agricultural development, particularly in the smallholder sector. Increased crop production amongst India/NDDB and Ethiopian smallholder dairy farmers are two of the many examples of this linkage.

It is imperative that the perception of failure that still surrounds livestock development be cast aside and substituted by a positive attitude that truly reflects the potential for livestock project implementation. Practical field experience can highlight both the pitfalls and opportunities for success.

Table: *Distribution of Performance Evaluations for Agriculture and Rural Development Operations - Evaluated During 1974–1988*

	S[a]	*U*[a]	*S (%)*
Area Development	68	61	55
Livestock	43	35 [b]	55
Irrigation	99	24	80
Agro-industries	31	10	76
Other [c]	174	45	79
Total	415	175	70

[a] S = Satisfactory; U = Unsatisfactory.

[b] 20 of the 35 unsatisfactory and only 7 of the satisfactory projects were inAfrica.

[c] Includes perennial crops, credit, fisheries, land settlement, research and extension, programme loan/credit and agricultural services projects.

With respect to field experience in the development of livestock production, there are innumerable factors which might affect the outcome of a production programme, many of which are country- or species-specific. This paper considers the following key areas of livestock project development.

- Appropriate Technical Packages;
- Credit and Risk Management;
- Adequate Incentive Framework;
- Adequate Supporting Services;
- Project Preparation & Implementation.

Appropriate Technical Packages

Getting the technology right is essential to successful project implementation. There is growing evidence that if technological innovation makes sense to farmers then extension services, even enhanced, play a marginal role in speeding the process of dissemination.

The frequent and ongoing failure of large ruminant breeding programmes is a classic example of inappropriate technology. The relationship between genetic merit, management inputs and productivity has been understood for years, yet Governments and planners continue to develop breed improvement projects with no conceived or agreed mechanism for capping exotic gene levels. A list of such examples would be almost endless and reflects a failure to interpret and transpose knowledge and research results from one production system to another.

Simple technologies which generate quick cash returns work best. Although that may appear to be a statement of the obvious, it is a principal that is rarely implemented. The common sense and practical experience required by project management and technical assistance to select and promote such technologies is an unfortunately rare commodity. Going a step further, livestock projects should focus on technologies that lead to spontaneous adoption.

Leguminous tree based farming systems in the Philippines and Indonesia, the production of seed from leguminous fodder in Thailand and Ethiopia, the undersowing of crops with leguminous fodder in Ethiopia, and the transference of dairy/draft heifers under the Indonesian transmigration programme are examples of such spontaneous adoption of technology. Occasionally, the introduction of appropriate legislation has also resulted in positive spontaneous technology adoption.

The restriction of hillside grazing as a resource protection measure has led to the large scale adoption of stall feeding and fodder production in Nigeria and Indonesia with subsidiary benefits in the emergence of a more productive herd structure and improved fertility transfer from animal to crop.

Research and pilot studies must clearly play a role in large ruminant project development but so must the "educated guess", provided it does not place resource poor farmers in a position of undue risk. In my opinion, too much time is presently lost in retesting technology (often under inappropriate or artificial management conditions) which could equally be evaluated and refined within the development process. I would cite the development of integrated forage/ livestock production systems in Thailand and Ethiopia as examples of where skilled intuition has resulted in the development process leading the research effort. It is in this context that the value of competent TA is brought to the fore; by understanding farmers' needs, production systems and the environment in which they exist, competent TA can and should short-circuit the development cycle and rapidly produce meaningful results on the ground. Success breeds success, and I am convinced that the positive, early results that frequently lead from such intuitive transference of technology helps capitalise on the high staff and farmer expectations, enthusiasm and receptiveness to change that prevails at project commencement and can do much to instill the confidence in the project team and its beneficiaries that is required to carry a project down the long road to successful achievement of goals.

To cite a specific example, I would regard sustainable forage production as the weakest in ruminant development in both rangeland and integrated crop/livestock production systems. Conventional approaches to fodder production must be re-examined. The traditional rangeland strategy of improving water supply in under-utilised range, while providing short term benefit, has tended to undermine traditional regulations of these resources, leading to exploitation in the longer term. In integrated crop/livestock production system, reliance is often placed on relatively expensive (per energy unit) and frequently subsidised agro-industrial byproducts to support incremental livestock production. Again, short term objectives are fulfilled, but with byproducts typically in limited supply such a strategy does not often support sustained livestock development.

Project management must focus on developing a sustainable forage base for incremental livestock production. Low input forage production systems, integrated where practicable into cropping cycles, efficient forage/feed utilisation and assured forage seed supplies, are essential components of a sustainable forage base. Historically, project-based forage development programmes have tended to use a limited number of species and strategies, frequently in competition with crop production; forage production and utilisation have been focused primarily on improved stock; little attention has been paid to the efficient utilisation of either existing or incremental forage production; forage seed production is frequently overlooked, the quantities required being too small to interest commercial or State sector production units; and the results of forage production programmes are rarely monitored.

Forage development can be accelerated by more innovative project management. Forage production strategies can be cheaply and easily tested on farmed and non-farmed areas; identifying those production opportunities requires a more informed, intuitive transfer of experience and technology between production systems. High input forage production (fertilizer, irrigation, etc.) may provide economically viable returns for intensive livestock production systems (e.g. dairying). However, the real need of the Third World livestock sector is typically a reliable dry season protein supply. Low cost legume production, exploiting every possible production niche, bypass protein and nitrogen treated crop residues are as yet largely unutilised strategies that can significantly increase productivity and reduce morbidity and mortality, particularly of immature stock.

Limited seed availability frequently constrains forage development; again, a more flexible approach to production is required. Forage seed production should be developed as a pre-project activity and be focused on smallholder based, "truthfully labelled" forage seed production under contract. This approach effectively bridges the gap in seed supply between the initial promotion of new forage techniques and their eventual commercialisation. On the latter point, it is worth noting the success of the Ethiopian Fourth Livestock Development Project, where smallholder based leguminous forage seed production has lifted seed availability from 2 tons per annum in 1987 to 80 tons in 1989–90 and a projected 200 tons in 1990–91. This project also amply demonstrates the opportunities that exist within smallholder production systems to exploit low cost forage production niches within crop production systems and on communally managed lands.

Risk Management

Investment in large ruminant production systems in the Third World, particularly by resource poor smallholder farmers, is inherently risk prone. Many smallholder livestock management decisions are based on a risk-minimisation strategy, yet livestock development projects frequently expose farmers to risk without reasonable protection or provide protection in the form of subsidies that inhibit the development of sustainable production systems. Risk can and must be minimised through normal commercial interventions. Sensible credit management with realistic appraisal and subsequent repayment schedules and the appropriate use of principal repayment moratoriums during the initial production period can greatly reduce risk. Commercial insurance is another very effective and greatly under-utilised mechanism of improving, at reasonable cost, farmer confidence to invest in large ruminant production.

Similarly, reliable and preferably commercialised animal health services, particularly in the form of assured vaccination and drug supply, help abate smallholder risk. Sound market information, reliable production technologies and assured feed supplies, preferably generated from resources within the farmers' control, also instil investment confidence.

Incentive Framework

Large ruminant development projects will not succeed without an appropriate incentive framework. Market regulation and subsidised livestock product importation have seriously eroded the effectiveness of numerous livestock projects. The failure to develop a sustainable dairy industry in several African countries is inexorably linked to subsidised importation of milk products. Subsidised grain production in West Africa has resulted in marginal crop production in good rangeland areas, thereby both displacing ruminant production and reducing livestock producer incentives to invest in fodder. As noted by de Haan . In many countries, livestock producer prices have been artificially depressed under a policy ensuring cheap urban livestock product supplies, while on the other hand inputs including health care, water and fodder are often provided free of charge, thereby displacing the livestock producer from the cash economy, decreasing the incentives to sell surplus stock and undermining the financial viability and sustainability of supporting services.

The importance of an adequate incentive framework to the successful implementation of large ruminant production programmes is illustrated by the case of China. Since the lifting of major policy restrictions (private livestock ownership, land tenure, price control) in 1979, beef production over the period 1979–1988 rose by 12.2% per annum and milk production by 15.4% per annum. The average annual growth in livestock production over the same period has been 9.3%, far higher than the 3.7% achieved for the period 1952–1970 and 2.1% from 1971–1978. Project implementors can and should influence the policy environment within which they operate. Project monitoring and evaluation programmes can be a valuable source of reliable data, upon which policy issues can be reviewed and revised. Regrettably, M & E is not often effectively addressed in project design and is indifferently resourced and applied following project commencement. Equally, policy considerations that might emanate from such M & E work often fall into the too sensitive/too difficult basket.

Support Services

Livestock projects are frequently developed as "islands of excellence" within a largely unproductive livestock sector, their successful implementation being dependent on non-sustainable, project-specific support, including credit and marketing services and unrealistically high manpower inputs for management and extension. While such an approach is convenient for the timely implementation of the project, its relevance to the development of sustainable production systems is questionable. Project designers and management must be more cognizant of project interventions that can lead to the development of sustainable, self-financing livestock production systems. Briefly, some considerations with respect to livestock project support services and their link to sustainable livestock development include:

- More attention must be paid to mobilising and strengthening traditional community linkages and social structures in support of livestock development. Effective consolidation of these resources enables the transfer of responsibility for project implementation from the public to the private sector. Farmer associations, built around traditional social structures, provide one of the most effective means for the management of natural resources and provision of basic extension and animal health services. In the longer term farmer associations may also offer an effective means of establishing a communication and planning interface with Government. Non-government organisations can play a valuable role in the constitution of such associations.
- Livestock projects rarely make adequate provision for staff training and particularly training that involves "hand-on" practical experience. Study tours and post-graduate studies provide useful experience but rarely equip staff with the skills required for practical implementation of improved technology. More emphasis is required on training programmes that take staff through a full production cycle and which involve practical field work, possibly as an assignment between successive theoretical training sessions.

- More emphasis is required on the commercialisation of livestock services with the public sector providing more of a catalytic, advisory and monitoring role. The privatization of veterinary and AI services, input supply and product marketing can lead to more efficient and sustainable services at the farm gate. In any event the provision of government services must not inhibit the development of parallel private services. Experience indicates that farmers are prepared to pay the full cost of these services provided their reliability and quality are assured.
- Livestock extension services often play only a marginal role in the adoption of new technology. The present emphasis on farmer/extension agent ratios and lines of communication is irrelevant without sensible technical innovation and accurate targeting of recipients. More emphasis should be placed on adaptive research and, as previously mentioned the educated guess, together with large scale field demonstration. Such inputs could be managed by fewer, more experienced technical staff working through community supported and directed farmer technicians. Extension staff should also be encouraged to participate in the production process. The Chinese contract crop production programme, although not without weaknesses, has dramatically increased productivity while rewarding both the farmer and his advisor. Similar pilot studies in the Chinese livestock sector, whereby the extension agent takes up shares in the farmers production programme thereby accepting both the risks and benefits of the production process, deserve close consideration.

Project Preparation and Implementation

Project Preparation: Development projects are unlikely to succeed without the establishment of a competent and motivated management team that fully understands and is sympathetic to the projects objectives and activities. Too often, this is not the case. The cause of this failure frequently starts with project preparation, in which national staff and particularly project implementors are inadequately involved. Development projects, which often dramatically alter resource availability and technical approaches, must be allowed to evolve slowly and be built on the foundation of a comprehensive understanding of the concerned sector.

Preferably this would be achieved through a sector review conducted in a collaborative manner by national staff and expert technical assistance. Such a review enables Governments to make educated decisions on development priorities and leads to the next step of project preparation involving the maximum level of participative planning with the beneficiaries. The use of the logical framework matrix or equivalent methodologies can significantly support this process. Such an approach is achievable though expensive. The expense, however, must be put in perspective.

The investment by Ethiopia of US$1.5 million between 1983–1985 in a livestock sub-sector review and the preparation of eight livestock investment projects led to a total investment in excess of US$100 million by international lending agencies. Preparation costs represented less than 1.5% of investment, modest by any standards. Furthermore, the extended and participative preparation process led to the establishment of committed project implementation teams who understood their task. Third world governments must be encouraged to invest in the project development process and must be supported in this process by international lending agencies, who too often take shortcuts in the pursuit of a larger loan portfolio. The major development financiers must further extend their forward planning systems to allow for sector studies, detailed evaluation of earlier investments, and appropriate preproject investment and initiation of procurement processes if they are to ensure early and successful project implementation.

Technical Assistance (TA) also plays an important role in project implementation. Technical competence, dedication and sheer hard work on the part of TA can do much to motivate and stimulate national staff. The success of a project should not hang on the input of the TA; nor should projects fail, as frequently happens, because of the lack of, or quite frequently the incompetence or disinterest of technical advisors. Appropriate evaluation criteria, personal interviews by project management of short listed candidates, the use of referees and increased, unbiased in-service evaluation, and increased emphasis on practical experience both by Third World governments and supporting development agencies would do much to relieve this persistent problem.

Project Management: The structure and composition of project management organisations are critical to successful project implementation. The use of an existing or modified sector management structure is preferable to the creation of project implementation units (PIUs). PIUs, even if shown to be answerable to sector management on the project organigram, inevitably take on an identify of their own with the result that the project is not longer perceived as an activity of the main Government-financed development effort. Project staff and programmes can as a result be sidelined by under-resourced and empire-conscious sector managers, thereby dramatically reducing the sustainability of project interventions. Comprehensive preparatory work opens the opportunity for sector rather than project loans, which in turn provide the opportunity to reorganise and revitalise moribund institutions and achieve truly sustainable livestock development.

As important as the management structure is the quality of the staff selected to manage and implement development interventions. There is no obvious formula to success here; however, the preparation of very clear terms of reference for management will help Government recognise the personal and technical qualities required and make management more accountable. The same applies to the technicians implementing development programmes, who must be drawn into the management decision making process. The establishment of forums which allow technicians to interact regularly with project management and the preparation of annual work plans by technicians for management review are essential features of successful project implementation. The establishment of these interactive processes should be conditional to loan effectiveness and closely monitored at supervision. Effective staff participation is a key to staff commitment and morale, which once lost is almost impossible to recapture.

Monitoring and Evaluation (M & E) is frequently omitted as an integral part of project preparation, being tacked on for inclusion through a project financed consultant input. This rarely works. Typically, by the time the consultant is in place and has developed and M & E format, a mass of critical data have been lost. Furthermore, development programmes once implemented do not take kindly to the superimposition of an M & E framework.

Another frequent weakness of monitoring is that it ends up as a one-way flow of information. Without beneficiary feedback, much of the value of monitoring is lost and the accuracy and efficiency of field data collection deteriorates. The increased adoption of project preparation methods such as the logical framework matrix would enable the more effective integration of M & E into project design. The danger of such methods, however, is that unless applied in a common sense manner, the methodology overwhelms its objective!

Project Supervision produces varying responses by project management. Generally, management feel threatened by the supervision process and consequently strive to present the project in its best light, studiously avoiding implementation issues. Regular, consistent and firm but flexible supervision can greatly improve project performance. Unfortunately, it is a rare commodity and project supervision is frequently indifferently applied. Project supervisors rarely have time for a thorough inspection of field activities and are frequently under intense headquarters pressure to increase commitment of funds. Furthermore, in the absence of an effective monitoring system, an enormous amount of project time and energy is devoted to preparing material for project supervisors, much of which is neither accurate nor relevant.

Regrettably, the UN system in particular has failed to adequately modify its methods of project implementation and supervision as it increasingly devolves project management responsibility to recipient governments. The World Bank approach of retroactive financing and half yearly in-country supervision contributes significantly to effective project implementation and deserves the close consideration of UN agencies, particularly FAO. Regular, consistent and flexible project supervision is certainly a cornerstone to successful project implementation. Greater attention to the objectives, form, timing and adequate funding of project supervision at the preparation stage would undoubtedly result in more effective project implementation.

Strategies for Sustainable Development of Small Ruminants

The majority (75.3 percent) of the mountainous regions of Africa are located in the east, notably in Ethiopia, Kenya and Tanzania. The high plateaux generally has good soil but high population densities. Agriculture is, therefore, intensive and permanent cultivation with little or no fallow is common. Crop and animal production are usually practised on the same farm, but in parallel rather than a real integration of activities (Jahnke, 1984). Despite these somewhat favourable condition, actual levels of subsistence are not much higher than in other tropical zones.

The greatest population densities in the high plateau areas exist in Rwanda and Burundi with an average of between 160–180 inhabitants per^2 km but with some areas in excess of 330 per km^2. Over 90 percent of these populations live in rural areas.

In Burundi, which will be taken as an example, agriculture is based on small farms (less than two hectares), supplemented by communal grazing land. For various reasons, these grazing lands are rapidly declining, both in area and in value, due to:

- the demographic explosion has led to population movements and settlement schemes are being implemented, notably in the plains of Imbo and Mosso which are traditionally semi-nomadic livestock herding areas;

- accelerated reafforestation programmes of ridges and numerous hills (several thousand hectares per year);
- intensified development programmes for certain crops (palm trees, rice, cotton); and
- over-grazing as a result of livestock being personal property while grazing is a communal resource and, therefore, there is a natural tendency of owners to increase livestock holdings to maximise their share of the increasingly limited grazing land.

In certain regions, population pressure is so intense that grazing land is disappearing. The country's future must inevitably lie in an expansion of agro-pastoral systems or, preferably, efficient agro-sylvo pastoral systems. Several such projects have been tried in recent years with relative success and are based on a synergy of food plants, cash crops and livestock production.

For subsistence, farmers plant various food crops according to altitude and climate and include: sweet potato, bananas, cassava, beans, maize, rice, eleusinia and "colocase" with surplus crops being sold. In addition, limited cash cropping includes: coffee, tea, oilpalm fruit and tobacco. Farmers may also keep a few head of livestock to serve as a marketable reserve and, at the same time, to produce manure for the crops. This latter function is often the primary reason for keeping livestock since breeding animals are rarely capable of producing sufficient milk for human consumption and little meat is required for home use. The manure is indispensable for assuring the survival and growth of plants.

Animal productivity, especially in ruminants, is generally low, due to the genetic quality of local breeds, poor nutrition (due to deteriorating rangelands) and animal health problems. Milk production is generally non-existent, fattening of the animals is very time consuming (also the manure produced is of poor quality) and is rarely undertaken.

Thus a multi-disciplinary approach is necessary to improve the agro-pastoral system. Efforts to increase productivity must be undertaken simultaneously with the provision of necessary inputs: seeds, improved breeding stock, processing and marketing facilities. Erosion control and improved supply of quality manure are also important objectives. When such systems are in action, the farmer will be able to produce sufficient for both subsistence as well as earning a supplementary income to raise their living standards and for farm improvements. Of course, such a system would necessitate thorough training of both extension staff and farmers for whom such systems may introduce new concepts and technologies.

Thus most existing projects aim at setting up efficient agro-pastoral systems based on the multitude of sedentary and diverse small farms. They address smallholders directly and aim at replacing semi-nomadic herds in favour of better managed smaller herds/flocks. The lack of adequate grazing land and intensified cultivation will mean that settling migratory herds, at least in the short-term, will necessitate a greater use of intensive housing. Thus, for animals to provide a sufficient source of income and manure, rapid improvements in genetic make-up, nutrition, reproductive performance and health must occur.

That such a system can lead to a reduction in cattle numbers is observed in Burundi where the cattle population has decreased from approximately 800,000 in 1977–1978 to around 400,000 in 1985–1986. The reduction in cattle numbers was associated with an increase in the number of small ruminants from 850,000 (1977/78) to 1,050,000 (1985/86).

In Burundi, goats (700,000 head) are the more prevalent than sheep (350,000 head). Sheep are sometimes ignored for reasons of prejudice but usually from a lack of interest; cattle raising is a preferred activity and under present production conditions the performance of sheep is mediocre. However, they are better suited than goats to the agro-pastoral systems now being established. This is due to their less selective and destructive feeding habits and their better adaptability to intensive management.

Strategies for Sustainable Development of Small Ruminants in Burundi

Before examining the possibilities for improved small ruminant production through disease control, management, feeding, genetics and marketing, it is essential that the existing small ruminant production systems are fully understood. It would be useless to promote interventions to improve productivity, unless they were adapted to the existing systems and are fully understood by the producers and fit in with their personal expectations. Problems in developing small ruminant production manifest themselves mainly in the areas of animal health, management, lack of technical skills, feeding, genetics and marketing. Therefore it was in these fields that applied research has to be conducted and solutions sought.

Animal Health

No action had as yet been taken to improve the situation regarding the health of small ruminants in Burundi. Epidemiological studies were non-existent and no prophylaxis programmes have been recommended, except for control of external parasites where dips and spray races already exist. Where such facilities do exist, producers usually only bring their animals for treatment at irregular intervals. This resulted in considerable losses, especially of young stock, estimated between 15 and 40 percent, although, the causes of morbidity and mortality differ considerably between regions.

Moniezia, Oestrus ovis, heartwater and plant poisoning (seasonal) can cause considerable problems at localised sites. Veterinary services, therefore, must draft, as precisely as possible, pathological profiles for the regions where they operate, to provide the basis for suitable prophylactic programmes. However, some afflictions such as helminthiasis are universal and the advantages of a regular worm-control programme are self-evident and permit a considerable reduction in mortalities at a reasonable cost.

However, experience showed that it was difficult to launch a development programme for small ruminant production without first identifying all the local pathological problems. Such activities would greatly increase the chances of success and avoid the discouragement of producers faced with diseases beyond their control.

All too often, the health aspects are neglected in projects when compared with the emphasis given to genetic improvement. It is essential to remember that changing an animal's genotype will not increase productivity unless it is in good health and properly nourished.

Improvement of the Farming System

As indicated in the introduction, this represents an essential factor in the development of small ruminant production in highly populated zones. Most projects have attempted to implement efficient agro-sylvo-pastoral systems which link agriculture, afforestation and animal production. In order to understand this type of study and project, one must consider the following:

- In the regions under consideration, fallow periods are short and usually limited to the first planting season (October-January).
- The topography of the land is very precipitous. Fields are cultivated on steep slopes which leads to sheet and gully erosion and inevitably soil loss. To limit this small terraces have been built with anti-erosion hedges of *Pennisetum purpureum, Setaria sphacelata* or *Tripsacum laxum*. At present these hedges are used to mulch coffee crops, but there is an increasing tendency to use them as livestock feed.

- An ambitious afforestation programme has been developed. The objective of the programme is to cover of 15 to 20 percent of the total land area with forest by the year 2000. These woodlands are being established both in large blocks and as small domestic plots. While this land is being lost to cattle raising it can, nevertheless, be utilised for sheep production.
- In the lower-lying foothills certain cash crops can be advantageously combined with sheep production eg. young oilpalm plantations (2–6 years) where nutritious leguminous fodder plants, such as, *Pueraria javanica, Desmodium uncinatum* and *Centrosema pubescens* plants are usually grown as a ground cover.
- Many crop residues and agricultural by-products (non-conventional foods) are found in the small, mixed farming systems of the highlands. Although no qualitative or quantitative inventory has yet been prepared, these might represent an appreciable nutritional resource for small ruminant producers.

Considering these factors, two improved farming systems are presently being implemented to improve small ruminants production:

An Integrated Small Ruminant/Crop Programme. This programme is aimed at establishing various mixed farming systems which include: the introduction of highly productive goats and sheep; the transformation of some fallow areas into forage plots where animals can either be tethered or herded; and the development of small farms for producing limited supplies of breeding females, weaned animals (for fattening) and high-quality manure.

The programme foresees a growing number of pilot farms, the owners of which must subscribe to certain conditions, notably:

- construction of pens for their animals .
- a regular supply of bedding
- planting of fodder crops:
  - o 300 metres of 80 cm wide anti-erosion hedges, comprising a mixture of *Setaria sphacelata* (or *Pennisetum purpureum*) and *Leucaena leucocephala,*
  - o 2000 m^2 parcel of a vetch-oat mixture in the first planting season, to be used as silage,
  - o following a health plan set by the project,
- acceptance of a breeding plan and basic record keeping.

A Programme Combining Sheep Production and Re-afforestation. This programme would have the following goals:

- to utilise the significant plant biomass in the under-storey of reforested areas which is also a fire risk,
- to diminish the cost of forest maintenance (labour, machinery or chemicals required for clearing the undergrowth), to benefit the afforestation programme with an additional supply of manure (the neighbouring crops would also benefit from the manure produced in the sheep pens at night (±500 kg/sheep/ year),
- to increase the profitability of the afforestation programmes through the sale of sheep, and
- to involve the rural population with the forestry projects through the establishment of associations of sheep breeders to collectively manage their flocks in the forest areas..

Feeding

The existing feeding practices depend on farming system and degree of intensification. Natural Pasture is the basis of the small ruminant diet, although utilisation differs between species. Goats constantly seek a varied diet of grasses, shrubs and forbs, while sheep prefer shorter grasses better suited to their labial morphologic structure - although at the same time increasing the danger of parasite infestation.

Natural pasture may be utilised by scavenging, tethering or herding. If well managed, the two latter methods permit a more rational use of the available biomass, combining good nutrition for the animal with adequate regeneration of the plant cover. In tree-crop plantations, natural vegetation can best be used by rotational grazing. In the pilot farms the predominant grasses were species of *Hyparrhenia, Eragrostis* and *Digitaria.*

Fallow land and roadsides are an important feed resource and which often differ in agrostologic composition and nutritional value from other natural pastures. The commonly occurring palatable species include: *Erlangia spissa, Bidens pilosa, Sorghum vulgare, Monathoxalis orophila, Guizatia scabra* and *Melinis minutiflora.* Cultivated pasture is occasionally used.

A diverse range of fodder grasses and tropical legumes have proven successful in long-term testing. Among the best-known fodder grasses are *Brachiaria ruziziesis, B. mutica, Setaria sphacelata, Panicum maximum, Cenchrus ciliaris, Cynodon plectystachyon* and *Digitaria unfulozi.* The better known legumes are *Centrosema pubescens, Desmodium uncinatum* and *Stylosantes guianensis.*

However, these pastures are expensive to establish and maintain and require costly fertilizer. In practice, only part of the available biomass is utilised and wastage may be in excess of 50 percent, however, controlled grazing can limit this wastage to 25 percent. Neither the species or the limited composition of cultivated swards make them suitable for goats. To sum up, the establishment of cultivated pastures can seldom be justified for small ruminant production.

Fodder Crops represent a further opportunity for feeding sheep and goats. The productivity of fodder crops, combined with the limited dry matter intake of small ruminants, allows satisfactory results to be obtained from a small cultivated area. Intensive fodder plots permit either year-round feeding using zero-grazing or tethering, or to assure sufficient feed during critical periods of the year. However, in densely populated regions with high crop intensities, available land is extremely limited for either cultivated pasture or fodder production, particularly during the first planting season.

Thus the following procedures have been recommended:

- During the first planting season, annual fodder plants should be planted and conserved as hay or silage. A mixture of vetch and oats is recommended: 40,000 – 50,000 kg/ha of this mixture produces a good leaf-stalk ratio and nutritional value.
- Perennial fodder crops, in the form of anti-erosion hedges, should be planted along the borders of terraces to minimise encroachment on cultivated land. These hedges should be composed of highly productive fodder grasses and legume trees, such as, *Leucaena leucocephala.*
- The advantage of *Leucaena leucocephala* is that it is highly nutritious, well accepted by small ruminants and wastage is usually low compared to other fodder types. Problems with mimosine toxicity can be overcome by inoculating specific bacteria into the rumen (if it does not exist) which will cause the complete breakdown of the toxic alkaloid mimosine-dehydroxy-pyridone.
- Occasionally perennial fodder crops can be planted on appropriated sites as the opportunity arise. For example, oilpalm plantations may incorporate cover crops such as *Puearia phaseoloides* and other legumes in the initial years.

Non Conventional Feeds (NFC) are supplementary feeds which should not be overlooked and can represent a significant percentage of small ruminant rations at certain periods of the year. Principle NFCs in the pilot farms include:

- Leaves, particularly appreciated by goats, include *Acacia* spp., banana, Leucaena, cassava and sweet potato vines, and
- crop residues left in the fields (stovers and straw) or by-products such as tops, husks, grape-stalks, etc.

Genetic Improvement

Genetic improvement may be achieved through either selection from within local breeds or by introducing exotic breeds for crossbreeding. However, even in those countries which have followed a policy of importation, priority is still given to studying of the potential of indigenous breeds which have show excellent adaptation to the local ecological conditions and which represent an indispensable genetic resource. Therefore, before ambitious crossbreeding programmes are undertaken, a careful selection of the best local breeding animals, principally male, would be advisable. Internal Selection within Local Breeds: Such a selection could include two aspects: culling and establishment of a reservoir of quality breeding animals, principally males.

- Culling. This aspect is of secondary importance since males are marketed early for either family consumption, to avoid either theft or conflict. While widespread castration of low quality males would probably lead to improved meat quality, it would not necessarily have a positive influence on offspring.
- Establishment of a Reservoir of Quality Breeding Animals.

Small ruminant flocks appear to lack good breeding males. Often producers rely on the free-ranging males from neighbouring farms to service their females. Moreover, they usually sell their young males quickly to avoid the risk of theft or conflicts resulting when females in heat are pursued over planted fields. The consequence is that negative selection takes places with the weakest and slowest growing males surviving and often only immature males being available to serve on-heat females. Even when a producer acquires a high-quality male it is usually sold immediately after the females become pregnant.

Therefore, the introduction of service stations is desirable. Such stations would be responsible for acquiring good quality bucks and rams and making them available on demand. Selection criteria should be based on phenotypic appearance plus whatever performance information is available eg. weaning weights.

Such service stations could later be responsible for taking some of the resulting offspring to initiate progeny testing. After a few years, this would identify selected local animals as the possible basis of a line-breeding programme. In order to facilitate collection, analysis and use of data, service stations would need to be associated with and academic or scientific institution.

Introduction of Foreign Breeds Through Cross-breeding: The choice of cross-breeding scheme depends on the objectives of the breeding programme:

- commercial cross-breeding using a terminal sire with the objective of producing animals for slaughter, or
- cross-breeding for stock improvement using appropriate techniques such as: back-crossing, rotational-crossing, or criss-crossing.

However, not all individual animals are suitable for crossbreeding and success is best assured when starting with a stock of local animals of recognised quality based on performance registration. A number of cross-breeds have been tested in the densely populated highland regions with varying degrees of success. Considering the reduction in cattle numbers and milk being a traditional component of the diet, local goats were initially crossed with milk goats including the Alpine, Anglo-Nubian and Saanen.

These crossbreeds were not however popular and their lean conformation was not appreciated by local goat-raisers. Furthermore, the market for goat milk is limited principally to the urban areas. The Boer breed, better suited for meat production, has been more favourably received. However, it remains difficult to obtain a sufficient supply of these animals for breeding purposes.

A few crossbreeding attempts have been undertaken with sheep. The Romney-Marsh was introduced to cross with indigenous breeds but was not a success and the programme has been discontinued. One reason is that the situation in Burundi is totally different to the highlands of neighbouring Kenya and Ethiopia were numerous exotic sheep breeds have been tested with some success, notably: the Merino, Corriedale, Hampshire, Romney-Marsh, Awassi and Dorper.

Marketing

Very little is known regarding the marketing of sheep and goats and systematic studies are required to determine the following factors:

- age, categories and sex of animals marketed,
- destination of animals sold,
- the role of the various intermediaries,
- the social status of the buyers,

- the size of the sellers' small ruminant stocks, and
- the primary reasons for the sale.

Extension Service

A general complaint of producers concerns the lack of technically qualified extension staff to assist in small ruminant production. Such services are practically non-existent at present and without them, farmers cannot see the potential for improvement. Not only do the extension services provide insufficient advice and training in the small ruminant production they also lack logistical support, infrastructure and even the simplest support materials.

It is interesting to note that those African countries that are able a policy to expand their small ruminant production (Kenya, Ethiopia, and the Ivory Coast) have also progressively developed their extension services. In Burundi significant efforts have been made in the area of training, both of extension staff and farmers.

The extension service should be responsible for addressing the following main constraints associated with sheep and goat development:

- the pathology concerning internal and external parasites and pneumonias etc.,
- migratory herding,
- inadequate feed and water,
- high levels of inbreeding and uncontrolled mating, and
- poor housing and insanitary conditions.

Training programmes need to be conducted at all levels:

- *National:* where a number of specialists should acquire an external specialised degrees,
- *Regional:* local extension agents with general backgrounds should be provided with further training through specialised courses, and,
- Producers: who should be offered a programme of short-term courses on specific subjects, such as, improved management, feeding, housing and care of the animal.

Results and Comments

Integrated Small Ruminant/Crop Programme

This project was well accepted by farmers, who were enthusiastic and the number of volunteers exceeds the project's capacity to provide fodder and animals.

At the outset, farmers fulfilled all conditions required for the loan of 5 ewes and/or does, although, thereafter, various features and impediments were identified. Apprehension regarding theft of animals induced farmers not to use pens but to bring their animals back to the family house.

A number are continuing to keep sheep and goats outside in pens but will usually guard them at night. In both cases the production of a large amount of well-rotten manure is definitely compromised.

Fodder Crops: This component was not satisfactory and supplementary feeding is irregular. Harvesting of fodder was not carried out properly or at the right time. Increase of fodder production did not match increases in flock size. Furthermore, most of the fodder crops are multi-purpose and are also used for mulch and thatch - these alternative uses require an advanced stage of lignification.

It is therefore important to make an inventory of all the alternative uses of fodder crops at the farm level and outline a chronological schedule that takes into account seasonal priorities and the required vegetative stage in order to fulfil both family and animal requirements. With planning it should be possible to develop cropping schedules that allow for the maximum production of fodder, using conservation techniques as appropriate, as well as satisfying the thatch, stake and mulch requirements of the farm.

Non Conventional Feeds: Available NCF are not fully utilised correctly. A survey was conducted in 1989–1990 in three similar projects to assess the quantity and quality of NCF at farm level, seasonal availability and possibilities of storage. Data collected is being analysed and a report is in preparation.

Special attention has been given to *Acantus spp.*, a natural thorny shrub widespread throughout the whole country and growing mainly on short-time follows and road sides. It has good nutritive value (±25% Crude Protein) and is well accepted by goats, but not by sheep. Farmers hesitate to use it because it is supposed to favour the expansion of echthyma.

Genetic Improvement: The programme is constrained by the lack of young males and negative selection where the fastest growing males are sold at a young age for slaughter and not kept for breeding. A need for good males, therefore has been identified and a programme has been started to purchase, breed and exchange young bucks and rams in Selection Centres located within the different projects involved in the National Network for Research and Development of small Ruminants. All performance data collected by the projects are collated in a data bank and analysed by the Department of Animal Science (University of Burundi) which co-ordinates the National Network.

Elite males, identified in one experimental site, are exchanged with other projects to avoid possible in-breeding.

Sustainability: Effective sustainability is a long term process and should assure the effective integration of the various components of the system. After four years experience with the National Network the following positive features can be highlighted:

- Increase of Cash Income: Sheep and goat sales have become the most important source of income for the farmer, even exceeding food and cash crops.
- Land Protection: The planting of fodder hedges (*Tripsacum, Setaria, Pennisetum, Hyparrhenia*) and fodder-trees (*Leucaena, Calliandra*) has considerably reduced erosion and restored soil fertility.
- Increase of Food-Crop Production: The use of a larger amount of a well-rotted compost, linked with the correct use of the pens restraining sheep and goats, has increased crop productivity.
- Non Conventional Feeds: The increased use of NFCs has maximised the utilisation of the all farm outputs.
- Improvement of Diet: The inclusion of high quality animal protein improves the human diet, although on an irregular basis since animals are usually only slaughtered or consumed for ceremonial purposes.

Outputs of the programme were mostly positive, however, a negative impact was identified concerning women's rights. Within the traditional system, only small numbers of goats and sheep were kept and some could be owned by women and children.

Subsequently, when the flocks increased and became an important part of the family income, men no longer accepted such sharing and claimed total ownership of all animals. Another negative aspect is the lack of motivation amongst extension staff.

The majority of the extension staff express little interest in small ruminant production and are, in any case, unskilled in the area. It would probably take a long time to build up an effective extension service at the smallholder level.

Combined Sheep Production/Reafforestation Programme

This programme has been conducted at three different sites:

- From 1981 to 1984 it was implemented in Rugazi (alt. 1200 m) in the Muwirwa Region which represents the transition area between the Zaire Nile Crest (alt. 2000–2200 m) and the Ruzizi (Imbo) Plain (alt. 800 m) in Central Burundi. It is therefore a very steep area with serious problems of soil erosion. A huge reafforestation programme was carried out between 1978 and 1984, when the European Development Fund withdraw its support for the programme. The woodlands used for sheep production were mainly softwoods of *Pinus patula* and *Pinus carribea.*
- Since 1985 the programme continued in Vyanda (alt. 1600 m) in the border area between the Muwirwa and Bututsi Regions in the southern part of the country with coniferous forest of *Pinus eliotti.*
- In 1987 a project was started in Ryarusera on the Zaire Nile Crest (2200 m) in eucalyptus forests.

All three projects were implemented in state-owned forests which had over-grazed common pastures. Since the first project has been abandoned, this paper will only discuss the last two projects which are funded by the World Bank. Research programmes examined potential management systems in order to maximise both outputs (timber and sheep) without detrimental effect on environment.

The results could be briefly summarised as follows:

- A programme has been prepared (until 2010) for the rational use of forest resources in order to assure sufficient grazing for the sheep flocks.
- The monitoring of vegetative cover under the trees has been established using agrostological and bromatological surveys every six months. This monitoring has been able to confirm the change in sward composition towards and more valuable grass species. This improvement is particularly significant in Ryarusera where interesting grass species have been identified i.e. *Panicum chionachne, Digitaria vestita, Hyparrhenia sp.*
- All growth and reproduction parameters of the flocks are collected and analysed at the Department of Animal Science (University of Burundi). Significant improvement in performance, related to the improved under-tree grass cover, has already demonstrated and current production as shown below:

	Vyanda	*Ryarusera*
Fertility (percent)	96.0	97.6
Prolificacy (percent)	111.0	130.0
Lambing interval (months)	8.8	8.0
Age at first lambing (months)	15.0	15.0
Mortality rate 0–12 months (percent)	15.0	6.9
Birth weight (kgs)		
Male	2.6	2.4
Female	2.4	2.2

The main objective of the project was to establish an experimental farm which, in addition, will also demonstrate to local sheep producers the benefits of combining sheep in reafforestation areas.

It will also allow producers to exploit progressively the state forest. To achieve this it is necessary to gather individual flocks together in communal flocks to facilitate and to simplify shepherding. Due to the individualistic nature of most local farmers it is extremely difficult to put this into practice.

Under-tree grazing is not sufficient to satisfy the feeding needs of sheep, especially during the dry season (May to October). Therefore, it is important to grow fodder crops, which can be established along the fire breaks, to supplement the diet. The main fodder crops/trees used are: *Tripsacum laxum, Setaria splendida* and *Leucaena diversifolia.*

Field Experience on the Development of Poultry Production

Sustainable Development" was defined by FAO and approved by the FAO Council in 1988 as follows:

> *"Sustainable Development is the management and conservation of the natural resource base, and the orientation of technological and institutional change in such a manner as to ensure the attainment and continued satisfaction of human needs for present and future generations."*

This definition contains two major components which are essential for sustainability. Firstly, the biological basis of sustainability (management and conservation of the natural resource base) and, secondly, the economic and socioeconomic aspect (continued satisfaction of human needs etc.). Consequently, this paper will address sustainability of poultry development in two parts:

- biological aspects, and
- economic and socioeconomic aspects.

Biological Aspects of Sustainable Poultry Production

The basis of sustainable agriculture, which is an important component of the global biological system, is the maintenance of the biological equilibrium as demonstrated in a simplified input-output-system. Assuming that energy is the crucial factor in the global bio-system, then the only continuous external source stems from solar radiation. Consequently, energy output must not exceed the supply potential for direct warming and biomass production.

The management of the environment (atmosphere, soil, water) which is the basis of production and metabolism of biomass is a second important factor. Products of metabolism, such as: heat, gases and minerals may be lost or recycled. These losses, however, not only change the balance encipher as they increase the demand for external resources on the input side, but may also damage the environment in the longer term. There is considerable potential for recycling of energy, nitrogen and minerals which can reduce these losses and help stabilise the biological equilibrium.

Taking livestock as a separate production system, it is evident that the energy balance is negative. It has been estimated that only 0.1 to 0.6 units of energy (in terms of consumable outputs) are produced per unit of energy on the input side (Zeddies, 1980).

This negative balance has to be made up by a surplus of energy retained by plant production. Therefore, livestock production may be sustainable from the biological point of view, as long as its negative balance is counteracted by surpluses in crop production based on inputs either from recycling or solar energy-mediated production.

In most developed countries, however, the positive balance of crop production in terms of renewable resources has decreased during the last decade (Pimementel *et al*, 1973) and is not large enough to cover the losses in animal production and the overall energy balance of the agricultural sector is negative. Estimates of the agricultural energy balance in Germany revealed that 1 unit of energy input produced only 0.36 to 0.28 units of usable/consumable products (Zeddies, 1980) - the balance was made up by high inputs of fossil fuel energy.

Considering that fossil energy reserves may only be available for a relatively short time, it is clear that the present agriculture production system is not biologically sustainable and, since livestock is an important contributor to the imbalance, the question must be asked can we afford to maintain livestock production at its present level. In addition to the negative energy balance, livestock have further negative effects on the environment, notably: methane and ammonia production which are known to aggravate/enhance the greenhouse effect and/or uncontrolled nitrogen and phosphorous outputs which could cause soil and water pollution.

The present situation may be justified because of the rapid growth of the human population and its need to feed itself. There is, however, no doubt that major changes must be initiated within the next decade unless a total breakdown of the agricultural system, and of livestock in particular, is to be avoided.

Examples and models do exist of farming systems, containing both crops and livestock, which have balanced or even a positive relationships between energy input and output (Nielson and Preston, 1981). Within these systems animal production plays an crucial role with regard to food supply and recycling of energy.

A highly efficient system, which makes maximum use of renewable resources, is a mixed farming model incorporating multi-purpose agricultural crops and animals (ruminants and monogastrics), a biogas digester and a fish point. While ruminants are essential for the efficient recycling of energy from fibrous crop residues; monogastrics are used to recycle waste products which are not suited for human consumption, such as by-products, insects, etc. The rest of the paper will demonstrate the relative efficiency of poultry production on different levels of intensity with regard to energy utilisation.

Extensive Versus Intensive Production Systems

In Africa and Asia more than 80% of rural farmers, even landless people, keep small flocks of poultry (chickens, ducks, guinea fowl and pigeons). These birds do not receive any regular feeding but survive through scavenging and obtain their feed from locally available natural resources. There is no doubt that this type of production is sustainable from the biological point of view since all the inputs stem from renewable resources.

The total production of such flocks is small and does not allow for any substantial off-take. The energy efficacy of such a bird for food production is very low and only approximately 4% of the feed energy is used for production. Since energy requirements for maintenance are covered by scavenging, any supplementary feed will be used for production thus increasing overall energy efficacy.

The production system will be sustainable from the puristic and biological point of view as long as the feed supplement derives from the positive energy balance of crop production. In many cases, however, fossil energy inputs are required to produce sufficient feed. It has been estimated that fossil energy accounts for approximately 20% and 30% of the total production inputs for grains and ready mixed feed (including transport, milling and mixing), respectively. Using these figures we can split the total energy inputs of poultry feed under extensive and intensive production systems into fossil and renewable energy. It becomes clear that only marginal inputs of fossil energy are required to produce progressive increases of egg production under improved traditional systems. Under intensive production systems, the share of fossil energy in the feed is more than 30%.

Additional inputs of fossil energy will be required for housing and equipment which will further increase the energy imbalance and reduce the sustainability of intensive poultry production.

While these figures may differ from country to country, the trend is clear and should be considered when poultry production strategies are developed.

Chapter 7

Technical and Research Considerations

Although change in socioeconomic factors is undoubtedly necessary to limit degradation, research has a clear contribution to make in defining problems and providing solutions, where and when socioeconomic conditions suggest that technical change may have an effect. Some of the topics that require research are set out below.

Socioeconomic Studies

Research, similar to Boutonnet's study of Algeria, is needed to define the socioeconomic situation in each country. This needs to make full analyses of feed supply from all sources, including any unofficial market and that grown illegally. It is also important to quantify trends in off-takes of slaughter animals to establish whether systems are becoming less efficient in most countries. Research is needed into the patterns of rights of access to rangeland: even if traditional systems have disappeared or been abandoned, it is too simplistic to assume that access is entirely free. The current situation, after a long period of ploughing in the steppes, should be described. It is clear that planting a barley crop establishes rights to the crop in its harvest year, but does it establish rights to the land subsequently? ICARDA is actively working on this in Syria.

Farmer Participatory Research (FPR)

Farmer participatory research is needed to define the farmers' view of their problems and discuss with them possible technical solutions. I see FPR as the only way to tackle the problem of low offtake of slaughter animals, which may be an increasing problem. Very close contact with flock owners is needed to identify specific causes and seek solutions.

Improvement of Extensive Grazing Land

The possibilities for improving productivity and sustainability of rangeland need investigating. Because productivity is limited by the amount of rainfall, technologies requiring costly inputs will never be economically viable in these arid areas. Options do exist and ICARDA has programmes on these for the WANA region.

In order of increasing input requirements, they are: Improved grazing management. In areas with annual plants, management directed to improving seed set will slowly increase productivity. Currently, these plant communities are grazed heavily in late winter, limiting early growth and making very little contribution to the animals' nutrient requirements. Lower grazing pressures and the deferring of intense grazing would slowly allow an increase in herbage production. For areas where the imbalance between animal numbers and pasture resources is not too great, guidelines for grazing management have to be established for flocks that are continuously shepherded. Traditionally high grazing pressures (animals/unit area/unit time) are used to achieve very high utilisation of the herbage available at a particular time. Continuous stocking systems are very difficult, if not impossible, to introduce for shepherded flocks.

The benefits of deferring the start of heavy grazing in early spring needs demonstrating, especially when shown that flocks walking long distances to graze very sparse pasture actually require more supplementation than continuously penned animals. Thompson (personal communication) showed that a flock taken to graze for 6 hours daily on poor pasture required 2.2 kg/head of supplement to maintain weight, while a continuously penned flock only required 1.7kg/head.

Application of fertilizer. In many soils in the region low soil phosphorous (P) limits growth and small applications of P in the presence of legumes can rapidly increase herbage production, if the grazing pressure is controlled. For example, legume seed numbers increased from 3.1 to 18.6 thousand per m^2 over 4 years with annual applications of 25 kg of P_2O_5 (Osman *et al*, 1990). Stocking rate was 2.5 times that of the unfertilized land.

Introduction of Legumes. Cocks (personal communication) suggests that, in Mediterranean areas, very small-seeded legumes of genera, such as *Astragalus* and *Trigonella*, may warrant research for areas with rainfall of about 250 mm. Species from these genera possess reproductive strategies adapted to these dry areas: very small seeds, that pass largely undigested through animals, large numbers of seeds per pod and high levels of hard seededness.

Introduction of Edible Shrubs and Trees. Shrubs, such as *Salsola vermiculata* and *Artemisia*, were an important component of natural steppe, not only as feed sources, but also to stabilise soils. Much research has been done on the establishment of plantations of bushes, particularly *Atriplex* species, both native and introduced, but there is a lack of long-term data on carrying capacity and animal performance. Although wide spread growing outside government stations is not common at present, they probably offer the only possibility for regeneration of degraded areas of steppe, particularly as low cost establishment from seed shows considerable promise.

Practical Technologies to Optimise Feed Resource Utilisation in Reference to the Needs of Animal Agriculture in Developing Countries

Production from a herd or flock of ruminants is a result of the interactions of environment, the animals nutrition, and its genotype. Individual productivity from most ruminants in all developing countries is low. The reasons for this are complex but in order of priority appear to be:

- the imbalance of nutrients that arise from digestion of the available forage resources when these are fed without supplements,
- the incidence of disease/parasitism, and
- the often harsh climatic conditions.

Genotype is over-emphasised as a primary constraint as poor nutrition has an overwhelming effect. Resistance to disease and high temperatures of particular breeds is however an important overall consideration.Recent nutritional research has demonstrated the possibility of very large increases in animal production that can be achieved by small alterations to the feed base. As these increases have also been achieved at the farm level without alteration to the other management practices, it demonstrates the large impact potential of the feeding strategies in the present environment.

Increased production of meat/milk with better body condition of animals also lifts lifetime reproduction rates. The lowering of the age at puberty of heifers and a decrease in the intercalving interval in cows have probably the greatest effect on the overall level of production.

Increased efficiency of utilisation of forage by the animals together with improved reproduction rates have demonstrated that production can be increased by up to five fold without changing the basal feed resources. This has been achieved, by providing the critical catalytic nutrients that are deficient in the diets and by balancing availability of nutrients closer to requirements. In general, the supplements required are urea/minerals and a source of bypass protein. In many countries these supplements are already available, in others, there is a need to manipulate these resources, to provide them in the correct amounts and in the appropriate form. In many grazing areas the basic resources are not available locally. Research and development is needed to produce them economically at the centres of ruminant population densities. In the rangelands, particularly in the semi arid areas, tree forages, seeds and pods represent by far the greatest potential source of protein meals.

Available Feed Resources for Ruminants in Developing Countries

Throughout the last thirty years the expansion of crop and livestock production in developing countries have more than doubled but the increase in demand for food has been even greater, leading to an increase of food imports by approximately 10% per year. This situation is expected to remain at the same rate for the foreseeable future.

The increased production of animal products in developing countries has been largely through increased animal numbers, while production per animal has been static or increased to a minor extent over a long period of time (Jackson, 1981). The demand for food for humans in these countries is likely to increase and, since cropping land is almost fully utilised at the present time, it appears that cultivated pasture is likely to become scarcer in most parts of the developing world.

The ruminants niche is likely to remain as a utiliser of carbohydrate biomass which is not digested by intestinal enzymes and, therefore requires fermentative digestion, and which cannot be used extensively by monogastric animals (i.e. forage, crop residues etc.). The ability of ruminants to convert otherwise waste biomass into meat, milk and other products and to accomplish work suggests that they will endure into the foreseeable future. The ever increasing pressure on land for crop production suggests, however, that they will have to continue to do this utilising crop residues, industrial by-products and pastures from relatively infertile rangelands. The common characteristics of such feeds are low digestibility, low protein content and a low mineral component.

Animal Productivity

Animal Productivity from Available Feed Resources: Research over the last twenty years clearly indicates that it has been a popular misconception that low productivity of ruminants in developing countries is a result of low energy density of the available forages (i.e. low digestibility). This concept, often repeated in reviews, even up to the present time, is misleading.

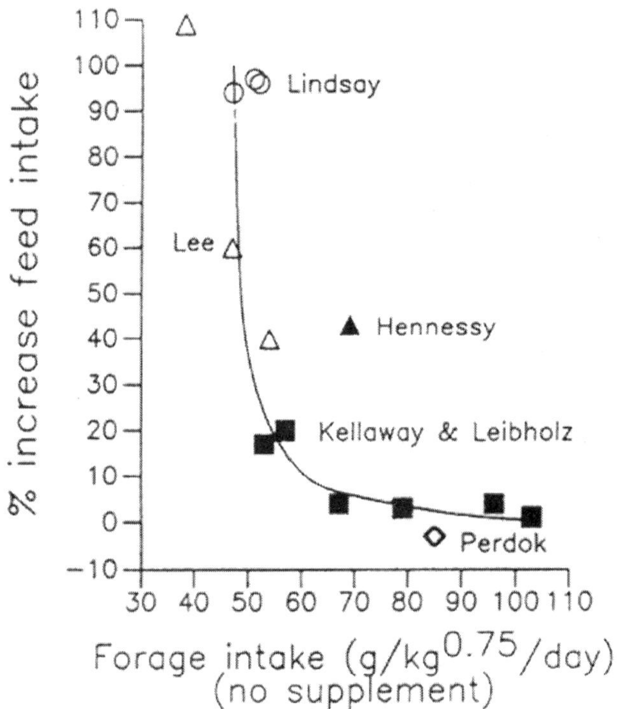

Figure: *Intake of low digestibility forages by cattle either unsupplemented or supplemented with bypass protein or bypass protein and urea (Lindsay and Loxton, 1981; Lindsay* et al., *1982; Hennessy, 1984, Perdok, 1987; Kellaway and Leibholz, 1981).*

There is now abundant evidence that low productivity stems from an inefficient utilisation of the feed because of deficiencies in the diet. These deficiencies are of nutrients critical to the well being of the microbes which ferment or digest the feed, and nutrients required to balance the products of digestion to requirements. This has been considered in detail recently by Leng (1991), who suggested that because an inefficient utilisation of nutrients increases metabolic heat, that the often low intakes of poor quality forage of ruminants in the tropics (where most developing countries exist) is imposed by a combination of climatic and metabolic heat stress (Leng, 1989).

Correction of a nutrient imbalance by feeding a bypass protein often (but not always) increases the intake of poor quality forages to a constant intake of around 80-100 g/kg$^{0.75}$/d. Tropical conditions impose this particular feed intake constraint on ruminants for a considerable period of a feed year and it is rarely seen in temperate areas. On the other hand cattle in the tropics require less feed for maintenance, if they do not have to combat cold stress, and if they can process these extra nutrients they can be more efficient than animals on the same feed in a cold climate. To process the extra nutrients however, they need extra protein and thus the requirements for amino acids are higher in cattle in the tropics then in animals on the same feed in cold environments.

Productivity Levels Achievable by Ruminants on Low Quality Forage Based Diets

The rationale and concepts on which the following discussions will be centred have been reviewed by Preston and Leng (1987) and Leng (1989,1991). The basic concepts are as follows. Ruminants fed low quality forages require supplementation with critically deficient nutrients to optimise productivity. The supplements that are required:

- correct any nutrient deficiencies for the rumen microbes, and
- they balance the ratios of protein (absorbed amino acid) to the energy (VFA) that arises from the fermentative digestion in the rumen so that it corresponds to the animals requirement.

The nature of anaerobic microbial fermentation in the rumen indicates that microbial cell growth in the rumen (which supplies the protein {P} to the animal) relative to the production of VFAs (E) (the major source of oxidisable substrate for ATP generation), will be extremely low in nutrient deficient rumen medium or digesta. This means that a sub-optimal level of any nutrient for microbial growth in the rumen will result in low protein to energy (P/E) ratio in the nutrients absorbed. Ensuring a nutrient non-limited microbial digestion in the rumen by supplementation automatically improves the P/E ratio in the nutrients available to the animal.

Feeding a protein meal in which the protein has been made insoluble or otherwise not attacked by rumen microbes, is a further major method for adjusting the P/E ratio upwards. The ratio of microbial proteins to VFA produced and the effects of supplementation in a steer given 4 kg of organic matter which is completely digested in the rumen for a number of feeding conditions.

The reason for discussing these theoretical calculations at this point is to emphasise the large differences in protein to energy (from 12 to 50) in the nutrients absorbed by ruminants, fed unbalanced diets and diets balanced with supplements. It also emphasises that on a diet high in bypass protein the rumen microbes need not be highly efficient. To obtain the higher P/E ratio without supplementation, however, to directly improve rumen condition, more bypass protein is needed. It is the relationship of P/E with the efficiency of feed utilisation that has a very large effect on growth, milk yield and reproductive performance.

The levels of production achieved when P/E is increased have been greatly superior to that predicted from present day feeding standards based on the metabolisable energy of a feed.

Table: *The effects on P/E ratio in the nutrients absorbed of supplementation with a bypass protein to cattle with a poor or optimised (i.e. supplemented) microbial milieu in the rumen. The values are calculated for a steer digesting 4 kg DM in the rumen.*

Rumen environment protein (g/d)	Protein bypass (g prot/d)	Microbial cells Produced	Protein microbial (g/d)	VFA produced (MJ)	P/E* (g.MJ)
Poor	0	830	500	41	12
Optimised	0	1680	1010	30	33
Poor**	400	830	500	41	22
Optimised	400	1680	1010	30	47

* *Microbial protein plus dietary protein to VFA energy.*

** *Although the rumen environment is deemed not to change through the addition of protein meal, in fact it will have been improved but may not be optimised to the extent it would by feeding a molasses/urea block. P/E ratio here is underestimated.*

Feeding Standards and Feed Evaluation

Most forages consumed by livestock in developing countries have a low digestibility which rarely exceeds 55% and is mostly in the range of 40-45%. The calculated metabolisable energy in the dry matter (M/D) thus ranges from 7.5 down to 4.8. Feeding standards indicate that feeds with a metabolisable energy content of 7.5 will support growth rates of cattle of approximately 2 g/MJ of M/E intake. On a forage at the lowest level of ME, cattle would be in negative energy balance.

The relationships found in practice with cattle fed on straw or ammoniated straw with increasing level of supplementation in Australia, Thailand (") (Wanapat *et al.*, 1986) and Bangladesh (Saadullah, 1984). Recent relationships developed for cattle fed silages supplemented with fish proteins (Olafsson and Gudmundsson, 1990)(Ä) and tropical pastures supplemented with cottonseed meal (Godoy and Chicco, 1990)(*) are also shown. This illustrates the marked differences that result when supplements high in protein are given to cattle on diets of low ME/kg DM.

Contrast this with results of supplementary feeding trials based on balancing the nutrition of animals with urea/minerals and bypass protein, where cattle growth rates equivalent to 18 g/MJ of M/E intake have been achieved in cattle fed straw. Obviously the presently accepted feeding standards have been very misleading and can not be used as a means of predicting animal performance. Of vital importance however, is that the application of the concept of balanced nutrition can improve animal growth by 2–3 fold and the efficiency of animal growth by as much as six fold over previous estimates (a range of 2–10 fold). In addition it also shows that although growth rates of cattle are below those on grain based diets cattle on forage based diets can be as efficient in converting feed to liveweight gain.

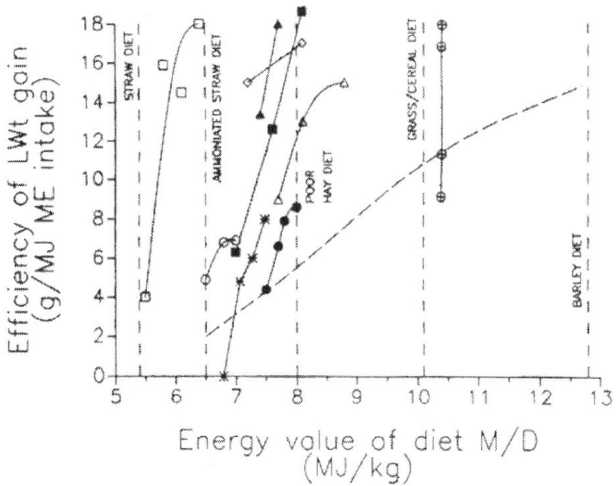

Figure: *Schematic relationship between diet quality (metabolisable energy/kg dry matter) and food conversion efficiency (g liveweight gain/MJ ME) from Webster, 1989.*

Low productivity of ruminant livestock has been accepted in developing countries as an inevitable result of the poor feed base and a low feed conversion efficiency.

The concept being that there is a large heat production (energy requirement) associated with the ingestion and movement of digesta along the tract in animals fed on forages as compared to concentrates. This conclusion is clearly contrary to the conclusions of Leng (1991) and the concept of balanced nutrition presented here. However, poor growth rates are associated with a slow maturity and, on poor quality roughage diets, cattle reach puberty at four to five years of age and often have a calving interval of 2 years. Recent observations on balancing the nutrition of cattle on such forages indicate that it is possible to reduce age at puberty by one to two years and potentially possible to support a calf every year on these same feed resources.

Management considerations, however, suggested that 15 months calving interval is more likely. The flow-on effects of improved reproduction are: increased percentage of a herd in production and an increased offtake of animals. The flow-on from improved reproduction outstrips the direct effects on immediate liveweight gain or milk yield.

Application of Supplementation to Balance P/E Ratios

Even though the principles of feeding bypass protein to improve productivity have been known for many years, application has been slow and unspectacular, particularly in the developing countries. The application has been slowed by:

- the inability of research scientists to communicate and be believed by applied technologists;
- the desire by many scientists to stand by the feeding standards that have been promulgated for twenty years and which appear to be totally inappropriate for most feeding systems;
- the controversies surrounding the principle mechanisms of action of protein supplementation which has clouded the major issues.

However, wide scale application of the use of supplements high in bypass protein have occurred in India through the initiatives of The National Dairy Board of India (NDDB). For the same reasons as given above, progress was initially slow (development started in 1980) but it is now accelerating at a pace which should see most feed mills in India dedicated to the production of bypass protein supplements in the next five years. At the present time approximately 100 MT of bypass protein feed is being fed daily to their dairy animals by small village farmers. In many situations, this is coupled with the use of a molasses urea block, particularly by the more advanced farmers. After considerable experimentation and village testing of a bypass protein supplements, the management of the cooperative feed mill in Kedah district in India decided to convert from the production of concentrates based on traditional concepts to a bypass protein concentrate (30% CP) composed of locally available protein meals plus 10% grain and approximately 10% molasses. After pelleting, the protein in the concentrate was found to be 75% insoluble in buffer solution (Leng and Kunju, 1988)

The marketing strategy was to widely advertise the new feed concentrate with a simple statement that village farmers should "feed to their dairy animals half the weight of the usual concentrate and this would double milk production". As the feed was about a third more expensive, there was some considerable opposition to its introduction. Nevertheless, the feed mill (capacity 100 MT/day) was converted to the production of new feeds on December 1st 1988. The feed was purchased, somewhat reluctantly by some of the farmers, but all opposition to its supply appeared to have been overcome by the middle of the next hot dry period (April-June) when milk yields were considerably above previous years with only half the weight of supplement.

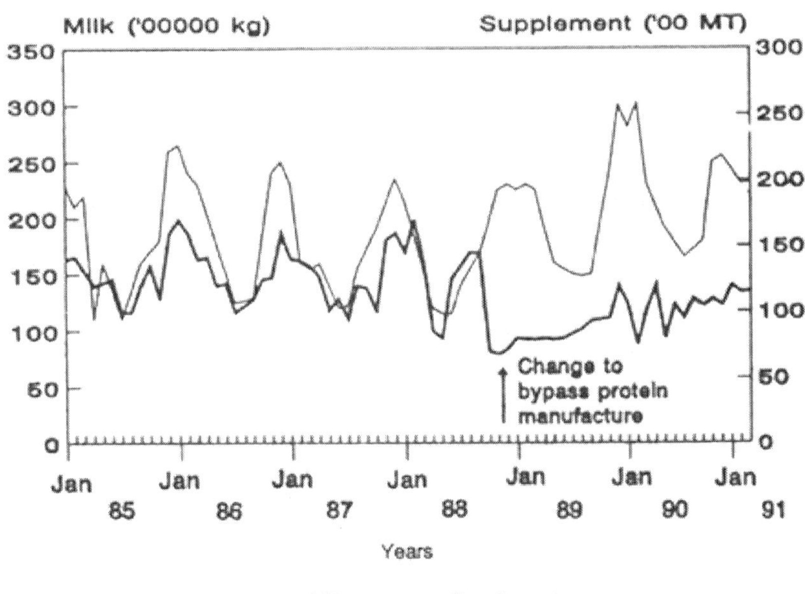

Figure: *Milk collection records and the sale of supplements in for the Co-operative in the Kedah district of India when supplements were compounded on traditional concepts (1985–1987) and following (1st December, 1988) their replacement with a 30% C.P. bypass protein pellet (records provided by the NDDB of India).*

The collection of milk within the dairy co-operative, relative to the feeding of the traditional and new feed supplement for the previous five years and for the twelve months since conversion to the new feeding systems. Whilst a number of changes have occurred in the area which could contribute to the increased milk collected, from the research carried out by NDDB prior to the change over the responses in milk yield are in line with observed increases in milk collected.

The effects of the changed feeding appear to be a 30–50% increase in milk production, from Dec. 1st. 1988 to Dec. 1st. 1989 - a further similar increase in production is apparent in 1989-1990 (unpublished observation). The further increase probably represents the flow-on effect that would come from increased reproduction rate that should have resulted from the new feeding systems.

Impact of New Feeding Systems on Milch Herd Composition

Undoubtedly the new feeding systems, combined with disease control and better management, is facilitating the introduction of cows with a higher genetic potential for milk production. The apparent lowering of heat stress through the balanced approach to nutrition has also allowed Friesians (the mothers of bulls that will be used for cross breeding purposes) and selected buffaloes, to yield milk well above the average reported for the developed countries. The yields of imported Friesian cows from Germany fed on all roughage diets supplemented with a molasses/urea block and a bypass protein (recommended at 350 g/litre of milk) at Anand averaged over 5,000 litres/305 day lactation in 1988–89 with individual milk yields of over 6,000 litres/305 day.

These milk yields are achieved in an area of India considered to have one of it's hottest climates. Higher yields were observed with Friesians managed by village farmers in a more temperate area of India (Bangalore).

Practical Methods for Balancing Nutrition of Ruminants Fed Low Quality Forages by Providing Deficient Minerals

Provision of Minerals

It is not practical to identify the major critical micro and macro minerals in a basal roughage diet, as these will vary from site to site and year to year. The practical approach is one of 'rules of thumb' that provide a best bet or 'shot-gun mixture' of minerals as economically as possible. A concentrated plant extract, such as, molasses provides such mixtures and can be fortified for specific areas where local knowledge points to specific deficiencies.

In this respect, molasses (both sugar cane and beet) and concentrated palm oil sludge offer useful sources of these minerals. They are also quite palatable to livestock and are useful in hiding less palatable nutrient sources in supplements.

Mineral salt mixtures are commercially available in most countries. They usually have a high content of salt and only minor quantities of the trace elements. In practice, fortified materials, such as molasses, will be superior to commercial mixtures as they present a greater coverage of all the minerals required, also thy are a valuable source of other nutrients (e.g. B vitamins) and a small amount fermentable energy.

Supplying the Rumen Microbes with Ammonia/Urea

The other requirement is for a non-protein nitrogen source for the rumen microbes, usually urea. Urea is usually administered together with the minerals and its concentration in such mixtures is controlled by safety and ease of incorporation and, therefore, rarely exceeds 10–15% of such mixtures. However, this is usually sufficient to allow an intake of between 50 and 100 g of urea by cattle from a molasses/urea block which is sufficient to balance ammonia in the rumen of cattle on a low N roughage diet. Results from India suggest that mineral/urea mixtures in, for example, molasses multi-nutrient blocks, are best given *ad libitum* allowing the animal some degree of selection and there are indications that the animal will learn to control the intake of urea to an optimal level.

Bypass Protein Supplements

Providing bypass protein to cattle under small farmer management is often difficult and, at times, is too expensive. There is often little information on the locally available protein sources, particularly the level of protection of the protein in the rumen.

As a rule-of-thumb, solvent extracted oilseed cakes, fish meal that has been flame dried (but not sundried or fish silage) and protein sources that have been heat treated, have considerable protection from rumen degradation. The degree of protection is enhanced by pelleting the protein meal in the presence of free glucose or fructose, (as occurs in molasses) when a mild browning reaction occurs (unpublished observations).

The turning over to supply bypass protein rather than concentrate for supplementing the cattle of small farmers, the National Dairy Development Board of India converted the existing feed mills to the production of high protein supplements. Wherever possible, solvent extracted oilseed meals form the basis of the supplement.

However, the protein ingredients have to be purchased on the open market and leastcost formulation is desirable because of the size of the production unit (50–100 MT/d.). The high degree of protection of these protein feeds is achieved by ensuring a high proportion of solvent extracted meal and also by heat pelleting with 8% molasses in the mixture.

Identification of Protein Sources

India is fortunate in having large amounts of crop residues high in protein, most of which have a fair degree of protection brought about by the processing methods. These materials are convenient to use in existing feed mills with the available equipment (i.e. grinders, mixers and pelleters) and the pelleted supplement is readily accepted because of a well developed marketing strategy.

In many other countries, particularly in extensive grasslands or savannahs, where major constraints to production of cattle are essentially the same as those for cattle fed crop residues, protein sources may not be readily available, or the sources not so obvious or easily obtainable. Most legume forages, legume seeds, edible tree leaves, seed pods and seeds that are available in these areas, contain highly soluble proteins which is easily fermented in the rumen.

These, when used as supplements, only provide ammonia and minerals (they have for example 0.5-1% phosphorus). Fed without processing, because of their influence in the rumen they generally increase production of cattle on a basal diet of low protein roughage, but they provide little bypass protein unless they are fed as a large proportion of the total diet. Research results in Colombia fit into the general concept where supplementation of cattle on green *Brachiaria decumbens* pasture with either urea/molasses or the foliage of the fodder tree *Glyricidia*, increased production to the same extent.

Table: *The effects of feed supplementation on livestock gain of cattle (6 per group) grazing on green* Brachiaria decumbens *pastures in the wet season (with mineral supplements) with liquid molasses/urea 10% or* Glyricidia *foliage.*

Treatment	Rumen Ammonia (gN/1)	Initial Wt (kg)	Final WT (kg)	LWt gain (g/d)
No supplement	50	194	244	580
+ *Glyricidia*	170	204	266	717
+ Molasses/Urea	250	203	269	751

(Source: ICA, 1988)

Protein that is fermented in the rumen yields 80–100 g of microbial protein/kg of protein consumed. Whereas carbohydrate yields 180–200 g of microbial protein/kg of carbohydrate. On a high nitrogen diet, a significant supplement of soluble protein could in fact imbalance the P/E ratio because of the low microbial growth efficiency.

This is a possible explanation for recent results reported from Iceland where fish silage and fish meal were compared as protein supplements to 'high quality' grass silage fed to young cattle. Fish silage, with the same N content as a fish meal supplement, depressed liveweight gain by 90g/day whereas a fish meal supplement improved liveweight gain by 360 g/day at the same N intake.

Table: *The effect of supplementation of a basal grass silage * diet with fish meal or fish silage on liveweight gain and efficiency of feed conversion (after Olafsson and Gudmundsson, 1990).*

Suppleent	Intake (kg/d) Supplement		Silage DM	LWt gain (kg/d)	Calculated ME intake	Efficiency (gLWt gain/MJ ME)
	DM	CP				
	.200	.123	4.6	.838	50	17
Fish	.400	.246	4.5	.914	51	18
silage	.400	.138	3.4	.358	40	9

* Early cut, precision chopped and preserved with formic acid, DM In vitro DMD. = 73% Calculated M/D = 10.4. CP = 16.8%. Cattle were Galloway/ Icelandic cross - 6 months (8/group treatment), 160–205 kg LWt at beginning of experiment.

The above discussion defines and highlights two potential strategies to provide the two types of supplement required to optimise the efficiency of feed utilisation of cattle in areas with scarce resources of nitrogen or protein. The strategies must be to find forages/trees/ tree seeds or pods that are high in protein and minerals. To then use this material in small supplements to provide for either the rumen with soluble protein and minerals or bypass protein after treatment to protect the protein. These uses can be combined with either a molasses/urea block and/or a source of locally available bypass protein.

Processing of Local Protein Resources to Provide Bypass Protein

The protein and render them non-fermentable in the rumen but allow them to retain digestibility in the intestines. In general, these fall into chemical treatment with agents that cross link with amino acids on the protein chain and include formaldehydes and aldehydes, tannins and simple sugars such as glucose and xylose. These reactions often require the protein source to be heated during processing.

Heat alone will often denature the protein to effect protection. For instance, Goering and Waldo (1974) found significant effects of the temperature used to dry lucerne on the subsequent animal production from that lucerne. In general, the higher the temperature of drying the greater the retention of protein by the animal.

More recently, Lewis *et al.* (1988) have demonstrated that mild heat with a small amount of xylose is very effective in protecting soya bean meal protein. Xylose can be readily produced by acid hydrolysis of many fibrous materials including bagasse and cottonseed hulls.

Table: *Effects on liveweight gain of cattle supplementing a basal forage/ concentrate based diet with soyabean meal or soyabean meal treated with sulfite liquor (SL) at 200°F for 2 hours (Lewis et al 1988).*

	LWt gain(g/d)
No supplement	591
+7% soyabean	673
+9% soyabean + 10% SL	823
+8% soyabean + 5% SL	841

Little work has been done with forage proteins, although there are indications that the presence of low levels of tannins in the forage (e.g. leucaena) does afford some protection of its protein. There is a great need to research methods for protecting the protein of forages, tree leaves, seeds and pods from local resources, as these are the proteins that are potentially available in the pasture areas of the world.

Trees as a Source of Protein and Soluble Nitrogen in the Rangelands

A number of authors have pointed to the small percentage of the total pasture biomass that is actually consumed by grazing animals in the extensive arid and semi-arid rangelands (Ellis and Swift, 1988). Most tropical grasslands are highly leached and their pastures apparently have low potential to provide ruminants with their required nutrients.

The point to be stressed is that 10–30% of pasture biomass is often all that is used by grazing animals. Trees, however, can produce considerable amounts of edible biomass. For example, the tree *Prosopis juliaflora* produces up to 440 kg of edible pods per annum with an average 16% crude protein, (Riveros pers. comm) but the protein is, in all probability, soluble. Compare this with the usable biomass of 250 kg DM from the poor native pastures of South America and the combination of trees and grassland would obviously be a desirable development and synergistic for cattle production.

The research needs are obvious. Local research in centres of cattle density must identify actual or potential protein sources; must then establish mechanisms for harvesting, processing to concentrate the protein if necessary and protect it from rumen degradation. Appropriate means for using both the processed and unprocessed protein/minerals to optimise the efficiency of animal production must be established.

Practical Technologies for the Optimal use of Tropical Pastures and Rangelands

Food production in some tropical regions has risen dramatically during the last two decades as a result of the green revolution, which consisted of intensified production with increased inputs, such as, fertilisers, improved cultivars and crop protection. The tropical pasture revolution, hailed in the 1960's, has also resulted in an increased potential for producing livestock products. However, production improvements have been less obvious than in the crop sector. This can be attributed to many factors, notably:

- there is both a greater need and market for staple foods, such as rice, and cash crops than for food of animal origin,
- livestock production is usually carried out in drier areas and on poorer soils, and
- in many countries animal products are regarded as luxury rather than staple food items.

Chapter 8

Future Challenges in Animal Production

If a sharply rising population in the developing countries is to achieve higher real incomes and a better quality of life, agricultural output must rise more rapidly than population growth. Further, if this rise in agricultural output is to be sustained over time, the natural resources which provide the basis for such output must be preserved and new technologies offering higher productivity must be developed. Increasing livestock production is an important component of this process, both because developing country consumers are expected to spend an increasing share of their rising incomes on livestock products, and because taking advantage of favourable livestock-crop production interactions is one approach to a more efficient, sustainable agriculture. Despite the generally favourable effect which the use of livestock has on agricultural resources, livestock are also a component of several unsustainable production systems. Among the most widely cited include:

- the overgrazing of arid and semi-arid lands, leading to range degradation and productivity decline,
- the destruction of humid rainforests for the establishment of pastures which degrade soils and quickly cease production,
- the pollution of watercourses from animal wastes resulting from intensive dairy and meat production, leading to reduced production and consumer welfare elsewhere,
- the production of methane gas as a result of ruminant livestock production contributing to the threat of a global "greenhouse" effect, and

- and the pollution of soils, watercourses and subterranean water supplies from the application of fertilizers, herbicides, and pesticides in the production of livestock feed grains.

Increased attention to the issue of sustainability in livestock production is important if these production systems are to be improved.

This paper is concerned with establishing a number of guidelines for optimising livestock production in developing or less developed countries (LDCs), consistent with achieving sustainable agricultural production. The paper focuses on economic factors (the author's specialisation) though it contains some thoughts on the role of political and social factors. The paper first provides an overview of recent achievements concerning livestock production in developing countries livestock production.

The paper then briefly discusses how different types of livestock are used as capital goods to produce multiple livestock products, and how the total amount of livestock capital and its use is highly responsive to the economic incentives provided to producers. The paper then analyses several of the more prominent types of government interventions in livestock markets, discussing their probable effects on economic welfare. An emphasis is placed on the need for research. Finally, several problems involving sustainability in livestock systems are discussed.

Recent Experience with Livestock in the Developing Countries

Within numerous national and international agencies, including FAO and the World Bank, concern has been expressed with the progress, or lack thereof, recently observed in livestock development. What does the record show? Have efforts to encourage livestock development been successful and what are the prospects for the near future? What policy changes could improve the situation?

Consider first the evidence regarding the rate of growth of livestock output in LDCs (Less Developed Countries), both absolutely and relative to population growth. One comprehensive study notes that food produced from livestock grew more rapidly than population for most products in Asia, North Africa/Middle East, Sub-Saharan Africa, and Latin America between 1961-65 to 1973-77 (Sarma and Yeung, 1985). These results, with the exception for eggs, held for meat in all regions apart from Sub-Saharan Africa, but for milk only in Latin America. On average, the output growth rate for meat, milk, and eggs in developing regions were 2.9 percent, 2.5 percent, and 5.3 percent, respectively. Thus, per capita meat output grew about 0.5 percent per year during this period, though there were enormous variations across countries. Per capita milk output remained stagnant whilst per capita egg production rose rapidly. This record must be considered mediocre, especially since the consumption of livestock products generally rose at a more rapid rate, leading to rising imports or falling exports.

A more recent study confirmed this result for milk (Jarvis and Saha, forthcoming). Milk output in developing countries grew at a steadily declining rate from 1960 to 1988, falling from 2.7 to 2.1 percent per year. However, the rate of output growth varied significantly across geographical regions and from one sub-period to another which suggests that milk production has been responsive to changes in the incentives faced by farmers. For example, although the growth rate of milk output was low and declining in Sub-Saharan Africa and Latin America (±2 percent) during the last decade where per capita incomes were declining; it was high and increasing in East Asia (nearly 3.5 percent in the last decade). The regional growth in milk output is highly correlated with regional economic growth, suggesting that demand for milk has been an important determinant of milk production. Indeed, growth was highest where the resource base (natural pastures) were limited and vice- versa. That milk output in the developed countries, which still accounts for 75% of world milk production, also declined during the last three decades. Nonetheless, subsidised exports from developed countries contributed to rising LDC milk imports and, in general, discouraged milk output.

Prices

Comparable data on real livestock prices are relatively difficult to obtain for a large number of countries, but it appears that, among meats in LDCs, beef prices have been relatively constant, again with wide variation across countries. Domestic prices have often been tied, at least loosely, to international beef prices which determine the cost of beef imports or the value of beef exports. International beef prices have remained essentially constant during the last three decades (Jarvis 1986). Within Latin America, for example, beef prices have risen in a number of major producing countries as consumption as risen more rapidly than production, forcing countries from a position as net exporters to self-sufficiency. These governments have gradually reduced interventions which depressed beef prices.

In contrast to beef, poultry prices have fallen steadily throughout the world, mainly in response to increased production efficiency resulting from: improved genetics, balanced feeds, veterinary drugs, economies of scale and better management. Poultry consumption has risen rapidly in response to its declining real price and is becoming the meat of choice in many LDCs, particularly in urban areas.

Milk prices have fluctuated widely in response to the variation in the level of export subsidies from developed countries, especially the European Community (EC). However, many LDCs have imposed tariffs or quotas on powdered milk imports to protect domestic producers. India, South Korea, Brazil, Chile and Colombia are examples and milk production has responded positively. However, as will be argued below, in each of these cases, technical change and increased producer efficiency has contributed more to the growth in output than have higher prices *per se*.

Indeed, milk output in each of these countries has been growing more rapidly than domestic consumption and, as each country has approached self sufficiency, governments have been reducing domestic prices. Future growth in the milk industry in these countries will depend on the growth of domestic milk demand and/or each country's ability to export milk in an increasingly competitive world market.

Return on Investment

What has been the profitability of livestock investments? In a World Bank ex-post evaluation of 104 agricultural development projects which were wholly or partially devoted to livestock, roughly 60% of the economic rates of return (EER) exceeded 10%, while 34% were below 5% (The World Bank, 1985). The average return on livestock investments, weighted by the amount invested per project, exceeded 12%. Although these are only moderate returns relative to those estimated for other Bank investments, livestock projects have offered an acceptable return.

More importantly, most of the projects which failed were in Africa, whose economic, technological and political context offered unusual challenges. More than two thirds of the projects in other regions performed satisfactorily. Analysis of these projects suggested that the variation in their performance was determined by many factors (The World Bank, 1985). Factors which had particularly strong effect included the availability or lack of:

- technological packages adequately adapted to existing farming systems,
- an economic context providing attractive producer incentives,
- an institutional capability for implementing the proposed project,

- qualified technical personnel to design and implement livestock development programs,
- a government commitment to livestock development, meaning a genuine willingness to establish a stable policy framework and commit the required public resources,
- political and economic stability, and
- the existence of clear property rights, which provide incentives for development and use of land and complementary resources.

Rather than suggest that there is something inherently difficult about livestock production, the World Bank's experience suggests that production depends strongly on the incentives faced by producers. These incentives, although strongly influenced by domestic income, population growth and by the international context within which countries trade, can be significantly affected by government policy.

Technical change is an essential component of any long term programme to increase livestock output, yet investment in livestock related research in developing countries has been low and the lack of appropriate technology has been an impediment the growth to livestock output. In summary, livestock output has been growing, but ruminant production (beef and milk) have been rising more slowly than monogastric poultry and pork. The trend in beef and milk prices has been relatively constant, whilst of poultry and pork prices have steadily declined. These figures suggest that technological change has been more rapid in the production of poultry and pork, allowing producers a favourable return while offering consumers a cheaper product. A higher level of technological change combined with protection of the resource base, is needed if ruminant output is to increase and provide consumers with more, cheaper and better products.

The uses of Livestock as Agricultural Capital

Livestock contribute in many ways to national welfare in developing countries. On average, livestock account for half of agricultural output when bothe their direct and indirect contributions are considered. Directly, livestock provide food and non-food products (hides and skins) amounting to about 20 percent of agricultural GDP. Indirectly, they contribute another 30 percent by supplying essential inputs to agricultural production. Livestock convert crop residues, agricultural by-products and pastures on marginal lands (resources with limited alternative use) into a range of higher value products for subsistence and sale.

The integration of livestock into cropping, via draught power and manure, increases the area cultivated, improves the timeliness of agricultural operations and helps maintain soil structure and fertility. As economic development proceeds, the use of ruminant livestock by small farms will shift away from the current emphasis on traction towards beef and/or milk production (Jarvis 1982, 1988).

Because animal food products command high prices, they are usually consumed in greater amount by individuals having higher incomes, although, in some countries in Africa and Latin America they may be account for a high proportion of expenditure even in poorer households. However, even where the poor consume few livestock products, it is economically attractive for them to produce such products and exchange these for other foods which provide cheaper sources of energy and protein. Both species and breeds vary in their capacity to produce different types of outputs and to utilise different types of inputs. Ruminants have the capacity to utilise low quality, bulky feeds such as pasture, crop and industrial by-products which have few alternative uses. Large ruminants (buffalo and cattle) also provide draught power and are by far the main source of milk.

Small ruminants (sheep and goats) are generally more prolific, produce wool and hair in addition to meat and milk and can prosper under poorer range conditions than large ruminants. Small ruminants are also a more convenient household source of meat, barter and cash,. They are also more easily stolen. In many situations, large and small ruminants are complementary since they utilise different forage species. Diversification of disease risks is further reason for running them jointly. Monogastrics (pigs and poultry) are utilised primarily for meat production, although hides, feathers, down and manure are also important products. The principal economic advantage of non-ruminants is their ability to convert high energy/protein feeds into meat at a more favourable ratio than ruminants. Such feeds are often expensive if they are demand in for direct human consumption. However, an increasing amount of high energy/protein feeds from crop by-products are now fed to non- ruminants and identification of cheap feed sources for monogastrics is an important aspect of livestock production in most developing countries.

In most countries there is a choice between two fundamental livestock production strategies a) the feeding of inexpensive, low-quality pasture resources to ruminants to produce meat, milk, wool, manure and draught power, and b) the feeding of high energy-high protein grains to non-ruminants for egg and meat production whenever the demand exceeds the amount which can be produced by ruminants from low-cost feed resources available. It is generally uneconomic to produce beef using feed grains, except where beef prices are unusually high, although the price of milk may justify such feeding.

Where pasture, forage or low-quality crop by-products are available (or can be economically increased), ruminants provide meat and milk at low cost. There is potential to increase pasture production in Latin America and so increase beef and milk production. A similar potential probably exists in much of Africa if trypanosomiasis can be controlled. In most other regions, pastures are limited and increased meat and milk production will have to come mainly from swine and poultry utilising grains and high-quality agro-industrial by-products. Such production will commonly take place in large, industrial-type enterprises close to urban centres. Despite the rapid growth in demand for poultry, smallholders in developing countries can be expected to rely on ruminants as their primary livestock assets because they use more efficiently the locally available, low quality feeds, and that they provide a wider range of products, particularly draught and manure, crucial to their overall farming system.

In meat production, smallholders can compete effectively with larger commercial enterprises only to the extent that they have access to low cost farm resources, especially feed and labour, which cannot economically be sold off-site. Such low cost resources usually result from the integration of agricultural and livestock activities. Many smallholders will find it profitable to maintain a small number of other species to utilise that available feed which ruminants do not utilise efficiently, to provide diversity to the family diet and to provide assets which can be liquidated in smaller amounts.

The Role of Government in Determining Livestock Production Incentives

Livestock are capital goods which are highly mobile and can be liquidated rapidly if economic incentives are unattractive. Livestock production can thus be strongly affected by government policy, both insofar as it affects:

a) the prices which farmers face for their products and for the inputs they purchase,

b) as it affects property rights (especially rights to land ownership and use),

c) the development of new technologies,

d) agricultural extension,

e) the availability and terms of credit,

f) animal health and sanitation, and g) infrastructure (e.g., roads, communications, and police and judicial services).

Although it is impossible in a brief paper to cover the workings of all policies in all countries, a number of the most important policies affecting beef and milk in developing regions during the last two decades will be briefly analysed.

As a general rule, economists believe that governments should work to ensure that the price paid to domestic producers of, say, beef or milk is equal to the international value of these products, i.e., the FOB value of exports and the CIF value of imports in a local port, plus or minus domestic transport costs to the site of production or consumption, respectively. The same rule is valid for establishing the cost of productive inputs. When this rule is not followed, national economic welfare is usually diminished, though there are some important exceptions.

Historically, many governments (e.g. Latin America) have tried to reduce the retail prices of beef in order to benefit consumers. However, lower prices to consumers usually require lower prices to producers, which lead eventually to lower output. Thus, the effort to assist consumers via lower prices is usually successful only in the short run, and may actually harm consumers in the longer run as output declines. Such policies have been adopted at least once by nearly every Latin American country during the last two decades. Although initially all consumers might have benefitted through the availability of beef at a lower price, this benefit would soon be lost for many as consumption is reduced or higher black market prices. Of course, producers lose in both the short and the longer term due to the lower price received. That the analysis is similar if prices are imposed at the producer or wholesale level.

A similar effect will be achieved if a country which exports beef imposes a beef export tax, reducing the domestic price of beef below the international level. The relationship between import and export on meat supply and demand where the import price of beef is higher than the export price. Given that the intersection of domestic supply and demand occurs at a price lower than the export price of beef (P_e), it will be profitable for the country to export beef. If the government allows free trade to occur, the domestic producer and consumer price will be equal to P_e, domestic consumption and production will equal C_o and Q_o, respectively, and exports will equal X_o.

Suppose now that the government imposes a beef export tax equal to t, where t is some fraction of the export value. The domestic price will decline to $P_e(1\text{-}t)$, causing an increase in domestic consumption, a decline in domestic production, and a decline in exports. Moreover, it can be shown that economic welfare falls by an amount equal to the two shaded triangles. These losses occur because consumers are now artificially induced to spend more on beef and because producers are discouraged from producing beef, thereby, diverting their resources into other activities which the country produces less efficiently. In addition, the reduction in the domestic beef price causes a significant redistribution of domestic income.

Beef consumers gain because the price of the product they purchase has declined; the aggregate gain for consumers is approximately indicated by the area abdc. In turn, producers lose because the price they receive for beef has also declined; their aggregate loss is approximately area aefc. The government earns tax revenue equal to the area tX_1. Because of the shift in income distribution, consumers may favour an export tax even though total national welfare is diminished by its effect. It may be noted that some countries which have imposed a beef export tax have also neglected to develop new beef production technologies, perhaps in the belief that the beef export surplus will remain. Over time, however, beef consumption in such countries may grow more rapidly than production (shown by a greater rightward shift in the demand curve than the supply curve).

If so, exports will decline. Once exports have been eliminated, the country will be simply self-sufficient and further increases in domestic demand will lead to a rising domestic price until the import price, P_m is reached. Subsequently, beef imports will begin.

Milk, like beef, is politically a highly sensitive product and has also frequently been subject to intervention. Such intervention has sought two distinctly different ends. In many countries, governments have concentrated on restraining prices. The effect has often been to depress output and decrease milk quality. For example, in Colombia the government has imposed price controls at various stages of the production and marketing chain for pasteurised milk, but not for crude milk and processed products. The expressed intent was to reduce the cost of pasteurised milk, thereby benefitting poor consumers.

The results have been somewhat different. Price controls on pasteurised milk have led to a) higher adulteration and lower milk quality, b) a higher percentage of milk marketed as raw milk with attendant health risks, and c) a lower price and higher consumption of processed milk products (cheese, yogurt, ice cream, etc.) which are consumed mostly by upper income groups, at the expense of a reduced supply of pasteurised milk, which is consumed more evenly by all income strata.

A different situation is one in which the government has imposed an import tariff on milk to protect domestic producers. This policy was commonly used by developing countries during the mid 1980s when international milk prices were strongly depressed by subsidised exports from developed countries. A number of countries which had rising milk imports (eg. South Korea, India and Indonesia), made significant efforts to develop their milk industry to satisfy a growing national demand.

Where the intersection of domestic supply and demand lies above the import price of milk. If free trade in milk (powder) is allowed, the domestic price will be equal to P_m, domestic consumption and production are, respectively, C_o and Q_o, and milk imports equal M_o. If an import tariff of t is applied, the domestic price will rise to $P_m(1+t)$, inducing a decline in domestic milk consumption, a rise in production, and a decline in milk imports. It can again be shown that national welfare is reduced by the area of the two shaded triangles. The rise in domestic price induces consumers to shift their expenditure to other goods whose contribution to consumer's utility is somewhat less than that which would have been offered by milk at the initially lower price. Similarly, producers are induced to shift more resources into milk production, which provides less benefit than could be achieved had these resources been used to produce goods in which this country was innately more productive.

That the import tariff which increases the domestic price again provides government revenue. It also increases producer incomes by an amount approximately equal to the area abdc, but reduces consumer welfare by an amount approximately equal to area aefc, which is larger than the gains achieved by producers and the government. Thus, milk producers are likely to favour an import tariff even if it decreases national welfare.

Although the conventional economic analysis laid out above indicates that the distortions introduced by a tariff on milk imports is likely to reduce domestic welfare, this analysis may be qualified in important ways. For example, if the international import price is temporarily low due to "dumping," but is expected to return quickly to a higher, normal level, a developing country may find it attractive to impose a countervailing import tariff to ensure that the domestic milk industry does not suffer irreversible harm during the period of low prices. An argument can be made that the gains from importing cheap milk during the interim period will be smaller than the long term losses inflicted on the dairy sector. Whether this is true depends on the relative costs and benefits of the alternative actions, which is an empirical issue likely to vary significantly from one country to another.

Similarly, if the country has potential to significantly increase its milk production efficiency in the long run, a transitory import tariff may permit, if combined with other actions like a strong research program, the realisation of such potential. This "infant industry" argument for a tariff is often advanced for the manufacturing sector.

A problem, however, is that the alleged increased efficiencies sought are often not achieved and the short run tariff becomes a long run tariff, protected by vested interests. If so, although producers gain from a higher domestic price, the loss by domestic consumers will be larger. In a number of countries, including India and Indonesia, domestic dairy development has been encouraged by significant import tariffs and/or the receipt of concessionary milk imports which were sold by the government at market prices.

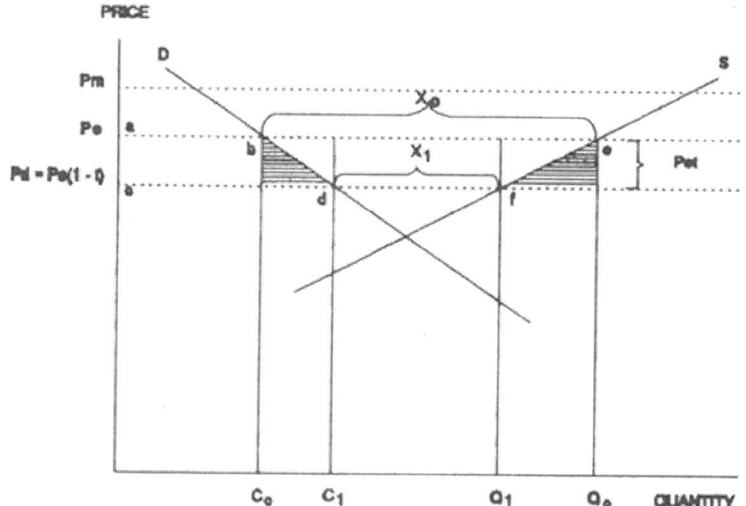

Figure: *The Welfare Effects of a Beef Export Tax*

The revenues thus obtained being used to subsidise the development of the domestic dairy sector (Alderman). Although such efforts have achieved significant improvements in the quantity and quality of milk produced, in each case, consumers have had to pay higher prices of milk than they would have had to pay had milk been imported at cheaper prices.

Thus, if cheap milk was desired for consumers, the policy of applying tariffs has achieved the reverse, at least in the short run. However, since higher income consumers account for the bulk of milk consumption, such policies might have distributional justification in some cases. Price intervention via export taxes, import tariffs, and fixed prices are common in developing countries. This paper has attempted to show that they can cause significant economic damage, even where they increase livestock output. The appropriate policy goal ought to be to optimise total economic output, not the output of any specific product - even livestock!

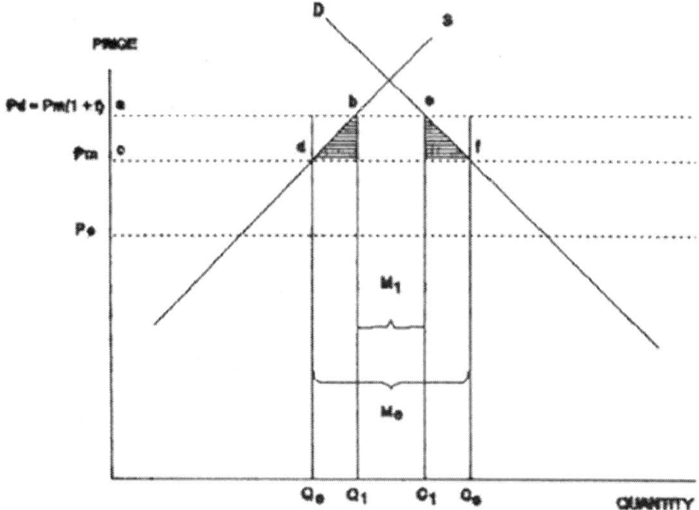

Figure: *The Welfare Effects of a Milk Import Tariff*

Several other types of economic interventions warrant mention. One is exchange rate overvaluation, which occurs when the domestic currency purchases more foreign currency than it would at an "equilibrium" in the balance of payments. Exchange rate overvaluation generally occurs as the result of economic development policies which restrict imports to encourage industrial development. The reduction of imports reduces the demand for foreign currency, which in turn leads to a stronger domestic currency. The stronger domestic currency directly reduces the domestic value of all traded commodities. For exported goods, for example, it has an effect similar to that of an export tax. It will do the same for imported products if these are not protected by tariffs.

Empirical studies of policies implemented in many countries suggest that the indirect effects of currency overvaluation have probably been more damaging to the livestock sector than have the direct interventions like price fixing and export taxes, which are more visible (Krueger, Schiff and Valdes, 1988). Rarely have agricultural sector spokespersons fully appreciated the damaging effects of currency overvaluation.

Even when they have, however, they have usually had little influence on such policies. Although agricultural interests usually influence the policies which directly affect the agricultural sector, the influence of such interests on macroeconomic policies, i.e., the exchange rate, is usually limited.

Government have occasionally sought to prohibit certain types of beef slaughter. For example, in Colombia in the mid-1960s and in Chile in the early 1970s, slaughter of heifers and cows that were deemed still "productive" was prohibited in an attempt to force expansion of the breeding herd and thereby output. This measure led to increased clandestine slaughter of females. Similarly, in Indonesia, the slaughter of younger bulls was prohibited with the intent of building up the livestock herd.

It can be shown, however, that the constraints imposed on producers by these prohibitions reduced overall livestock profitability by impeding the producer from taking what appeared the most profitable action (Jarvis 1982). Farmers generally perceive better the appropriate use of livestock capital than do governments. In each of these cases, when farmers slaughter animals they do not generally cease to own livestock assets, but simply shift from ownership of an older less efficient animal to a younger more efficient animal. The government's effort to dictate the appropriate use of animals thus doubtlessly led to a long-run decline in the livestock herd and in output rather than the increases sought.

In most developing countries, industrial protection policies (import tariffs and quotas) has increased the cost of inputs important to livestock production, such as machinery and equipment, veterinary supplies, minerals supplements, fertilizer, and fencing materials are usually high relative to costs in developed countries.

The lack of competition among importers of many brand name products is also a factor in high input costs. Such costs inhibit the development and adoption of improved livestock technology. Given the tendency of government market intervention to cause harm, most economists argue that governments should minimise such interventions. However, there are other non-market interventions, like those in public health, veterinary services, research and extension and the provision of infrastructure, which governments must undertake.

Livestock Research, Infrastructure, and other Government Interventions

It has become increasingly clear that achieving higher rates of livestock output associated with higher incomes for producers and lower prices and higher quality products for consumers, depends on obtaining improved technology. Probably no aspect of government policy is as important as research for the future of livestock production in developing countries. In several countries where efforts have been made to develop a domestic dairy industry, via import protection, the increase in output has probably come more from technological diffusion than from the stimulus of higher prices, *per se*.

For example, in India, the increasing use of crop by-products as feed concentrates, along with the introduction of crossbred dairy cattle having higher potential yields, has resulted in a rapid increase in milk output per animal. Similarly in Colombia and Brazil, the spread of pasture raised crossbred dual purpose cattle has led to rising output through a combination of increased herd size and higher output per animal. In each case, the output expansion has occurred while prices have been roughly constant or declining. Further, as domestic output has risen, reaching self-sufficiency levels, policy makers have been confronted with the possible need to export milk surpluses at world prices which are substantially lower than either the protected domestic price or, even, the milk import price. Even when basic technology is available in developed countries, a strong domestic research capability is essential to identify and adapt promising technologies to local conditions. Unfortunately, most livestock research and development institutions are weak and a major effort is needed to strengthen them, especially, the number of qualified staff.

An institutional environment which allows farmers to a) market their produce for higher prices, b) obtain inputs at lower cost, and c) obtain information necessary for decision marking more rapidly and effectively is critical to livestock development. Infrastructure is especially important where timely access to inputs is essential and where livestock product must be marketed quickly. Milk production is such an example and depends heavily on an efficient the establishment of a marketing and distribution system, including milk collection, transport, and processing. The diffusion of new technologies to farmers, e.g. artificial insemination and forage technologies, has often been achieved through the same systems. Such efforts have been carried out efficiently by private firms and cooperatives, though rarely by public firms.

Livestock diseases impose significant economic losses and can present a threat to human health. Such losses and threats can be economically reduced by improved animal health programmes. However, programmes such as vaccination, sanitation and inspection programs require a deep commitment to implementation from producers and governments and often require regional cooperation.

The Effect of Social and Political Factors

Social factors can influence the use of livestock such influences are often obvious, for example, religious objections to the consumption of pork in Muslim countries and beef in Hindu countries. In other cases, however, the effects of social influences are less pronounced though still important. For example, the attribution of prestige to ownership of cattle may increase the cattle's total value but lead to other tangible products like meat and milk. In general, such effects seem relatively small in magnitude.

In much of Asia cattle, utilising locally available residues, are used mainly for animal traction on small farms. In India, for example, cattle are fed principally on wheat and rice straws which account for approximately half the plant's total gross energy. This energy would be largely wasted if it were not consumed by ruminants and converted into traction for ploughing and manure.

Due to religious beliefs beef in India is of little value (beef is consumed by the Moslem minority in some areas). Nonetheless, farmers require some means to cultivate and most farms are too small to justify mechanisation and produce forage. The value from draught and manure must therefore be sufficient to justify maintaining cattle and this in fact the case where livestock are kept. Incorporation of manure into fields and garden areas, to improve soil fertility and structure, is often essential to maintaining agricultural production. It can be shown that the social constraint on the eating of beef, by diminishing the value of beef which is a necessary by-product of milk, traction, and manure production, makes cattle a less profitable investment. Further, with no incentive to slaughter the animal, the tendency is for producers to simply abandon their animals when their economic life animals has ended. Such animals are a general nuisance and harm other farmers who need to protect their crops and from consuming valuable common grazing.

Chapter 9

Production in Dairy Industry

A dairy is a facility for the extraction and processing of animal milk—mostly from cows or goats, but also from buffalo, sheep, horses or camels —for human consumption. Typically it is a farm (dairy farm) or section of a farm that is concerned with the production of milk, butter and cheese.

Terminology differs slightly between countries. In particular, in the U.S. a dairy can also be a facility that processes, distributes and sells dairy products, or a room, building or establishment where milk is kept and butter or cheese is made. In New Zealand English a dairy means a corner convenience store, or superette—and dairy factory is the term for what is elsewhere called a dairy.

As an attributive, the word dairy refers to milk-based products, derivatives and processes, and the animals and workers involved in their production: for example dairy cattle, dairy goat. A dairy farm produces milk and a dairy factory processes it into a variety of dairy products. These establishments constitute the dairy industry, a component of the food industry.

History

Milk producing animals have been domesticated for thousands of years. Initially, they were part of the subsistence farming that nomads engaged in. As the community moved about the country, their animals accompanied them. Protecting and feeding the animals were a big part of the symbiotic relationship between the animals and the herders. In the more recent past, people in agricultural societies owned dairy animals that they milked for domestic and local (village) consumption, a typical example of a cottage industry. The animals might serve multiple purposes (for example, as a draught animal for pulling a plough as a youngster, and at the end of its useful life as meat).

In this case the animals were normally milked by hand and the herd size was quite small, so that all of the animals could be milked in less than an hour—about 10 per milker. These tasks were performed by a dairymaid (dairywoman) or dairyman. The word dairy harkens back to Middle English dayerie, deyerie, from deye (female servant or dairymaid) and further back to Old English dæge (kneader of bread).

With industrialisation and urbanisation, the supply of milk became a commercial industry, with specialised breeds of cattle being developed for dairy, as distinct from beef or draught animals. Initially, more people were employed as milkers, but it soon turned to mechanisation with machines designed to do the milking.

Historically, the milking and the processing took place close together in space and time: on a dairy farm. People milked the animals by hand; on farms where only small numbers are kept, hand-milking may still be practiced. Hand-milking is accomplished by grasping the teats (often pronounced tit or tits) in the hand and expressing milk either by squeezing the fingers progressively, from the udder end to the tip, or by squeezing the teat between thumb and index finger, then moving the hand downward from udder towards the end of the teat. The action of the hand or fingers is designed to close off the milk duct at the udder (upper) end and, by the movement of the fingers, close the duct progressively to the tip to express the trapped milk. Each half or quarter of the udder is emptied one milk-duct capacity at a time.

The stripping action is repeated, using both hands for speed. Both methods result in the milk that was trapped in the milk duct being squirted out the end into a bucket that is supported between the knees (or rests on the ground) of the milker, who usually sits on a low stool.

Traditionally the cow, or cows, would stand in the field or paddock while being milked. Young stock, heifers, would have to be trained to remain still to be milked. In many countries, the cows were tethered to a post and milked. The problem with this method is that it relies on quiet, tractable beasts, because the hind end of the cow is not restrained.

In 1937, it was found that bovine somatotropin (bST or bovine growth hormone) would increase the yield of milk. Monsanto Company developed a synthetic (recombinant) version of this hormone (rBST). In February 1994, rBST was approved by the Food and Drug Administration (FDA) for use in the U.S. It has become common in the U.S., but not elsewhere, to inject it into milch kine dairy cows to increase their production by up to 15%. However, there are claims that this practice can have negative consequences for the animals themselves.

A European Union scientific commission was asked to report on the incidence of mastitis and other disorders in dairy cows, and on other aspects of the welfare of dairy cows. The commission's statement, subsequently adopted by the European Union, stated that the use of rBST substantially increased health problems with cows, including foot problems, mastitis and injection site reactions, impinged on the welfare of the animals and caused reproductive disorders.

The report concluded that on the basis of the health and welfare of the animals, rBST should not be used. Health Canada prohibited the sale of rBST in 1999; the recommendations of external committees were that, despite not finding a significant health risk to humans, the drug presented a threat to animal health and, for this reason, could not be sold in Canada.

Structure of the Industry

While most countries produce their own milk products, the structure of the dairy industry varies in different parts of the world. In major milk-producing countries most milk is distributed through wholesale markets. In Ireland and Australia, for example, farmers' co-operatives own many of the large-scale processors, while in the United States many farmers and processors do business through individual contracts.

In the United States, the country's 196 farmers' cooperatives sold 86% of milk in the U.S. in 2002, with five cooperatives accounting for half that. This was down from 2,300 cooperatives in the 1940s. In developing countries, the past practice of farmers marketing milk in their own neighbourhoods are changing rapidly. Notable developments include considerable foreign investment in the dairy industry and a growing role for dairy cooperatives. Output of milk is growing rapidly in such countries and presents a major source of income growth for many farmers.

As in many other branches of the food industry, dairy processing in the major dairy producing countries has become increasingly concentrated, with fewer but larger and more efficient plants operated by fewer workers. This is notably the case in the United States, Europe, Australia and New Zealand. In 2009, charges of anti-trust violations have been made against major dairy industry players in the United States.

Government intervention in milk markets was common in the 20th century. A limited anti-trust exemption was created for U.S. dairy cooperatives by the Capper-Volstead Act of 1922. In the 1930s, some U.S. states adopted price controls, and Federal Milk Marketing Orders started under the Agricultural Marketing Agreement Act of 1937 and continue in the 2000s. The Federal Milk Price Support Program began in 1949. The Northeast Dairy Compact regulated wholesale milk prices in New England from 1997 to 2001.

Plants producing liquid milk and products with short shelf life, such as yogurts, creams and soft cheeses, tend to be located on the outskirts of urban centres close to consumer markets. Plants manufacturing items with longer shelf life, such as butter, milk powders, cheese and whey powders, tend to be situated in rural areas closer to the milk supply. Most large processing plants tend to specialise in a limited range of products. Exceptionally, however, large plants producing a wide range of products are still common in Eastern Europe, a holdover from the former centralized, supply-driven concept of the market.

As processing plants grow fewer and larger, they tend to acquire bigger, more automated and more efficient equipment. While this technological tendency keeps manufacturing costs lower, the need for long-distance transportation often increases the environmental impact. Milk production is irregular, depending on cow biology. Producers must adjust the mix of milk which is sold in liquid form vs. processed foods (such as butter and cheese) depending on changing supply and demand.

Operation of the Dairy Farm

When it became necessary to milk larger numbers of cows, the cows would be brought to a shed or barn that was set up with bails (stalls) where the cows could be confined while they were milked. One person could milk more cows this way, as many as 20 for a skilled worker. But having cows standing about in the yard and shed waiting to be milked is not good for the cow, as she needs as much time in the paddock grazing as is possible. It is usual to restrict the twice-daily milking to a maximum of an hour and a half each time. It makes no difference whether one milks 10 or 1000 cows, the milking time should not exceed a total of about three hours each day for any cow. As herd sizes increased there was more need to have efficient milking machines, sheds, milk-storage facilities (vats), bulk-milk transport and shed cleaning capabilities and the means of getting cows from paddock to shed and back.

Farmers found that cows would abandon their grazing area and walk towards the milking area when the time came for milking. This is not surprising as, in the flush of the milking season, cows presumably get very uncomfortable with udders engorged with milk, and the place of relief for them is the milking shed. As herd numbers increased so did the problems of animal health. In New Zealand two approaches to this problem have been used. The first was improved veterinary medicines (and the government regulation of the medicines) that the farmer could use. The other was the creation of veterinary clubs where groups of farmers would employ a veterinarian (vet) full-time and share those services throughout the year. It was in the vet's interest to keep the animals healthy and reduce the number of calls from farmers, rather than to ensure that the farmer needed to call for service and pay regularly.

Most dairy farmers milk their cows with absolute regularity at a minimum of twice a day, with some high-producing herds milking up to four times a day to lessen the weight of large volumes of milk in the udder of the cow. This daily milking routine goes on for about 300 to 320 days per year that the cow stays in milk.

Some small herds are milked once a day for about the last 20 days of the production cycle but this is not usual for large herds. If a cow is left unmilked just once she is likely to reduce milk-production almost immediately and the rest of the season may see her dried off (giving no milk) and still consuming feed for no production. However, once-a-day milking is now being practised more widely in New Zealand for profit and lifestyle reasons.

This is effective because the fall in milk yield is at least partially offset by labour and cost savings from milking once per day. This compares to some intensive farm systems in the United States that milk three or more times per day due to higher milk yields per cow and lower marginal labour costs. Farmers who are contracted to supply liquid milk for human consumption (as opposed to milk for processing into butter, cheese, and so on—see milk) often have to manage their herd so that the contracted number of cows are in milk the year round, or the required minimum milk output is maintained. This is done by mating cows outside their natural mating time so that the period when each cow in the herd is giving maximum production is in rotation throughout the year.

Northern hemisphere farmers who keep cows in barns almost all the year usually manage their herds to give continuous production of milk so that they get paid all year round.

In the southern hemisphere the cooperative dairying systems allow for two months on no productivity because their systems are designed to take advantage of maximum grass and milk production in the spring and because the milk processing plants pay bonuses in the dry (winter) season to carry the farmers through the mid-winter break from milking. It also means that cows have a rest from milk production when they are most heavily pregnant. Some year-round milk farms are penalised financially for over-production at any time in the year by being unable to sell their overproduction at current prices.

Dairy Farming

Dairy farming is a class of agricultural, or an animal husbandry, enterprise, for long-term production of milk, usually from dairy cows but also from goats and sheep, which may be either processed on-site or transported to a dairy factory for processing and eventual retail sale. Most dairy farms sell the male calves born by their cows, usually for veal production, or breeding depending on quality of the bull calf, rather than raising non-milk-producing stock. Many dairy farms also grow their own feed, typically including corn, alfalfa, and hay. This is fed directly to the cows, or is stored as silage for use during the winter season. Additional dietary supplements are often added to the feed to increase quality milk production.

History

Dairy farming has been part of agriculture for thousands of years. Historically it has been one part of small, diverse farms. In the last century or so larger farms doing only dairy production have emerged. Large scale dairy farming is only viable where either a large amount of milk is required for production of more durable dairy products such as cheese, or there is a substantial market of people with cash to buy milk, but no cows of their own.

Hand Milking

Centralized dairy farming as we understand it primarily developed around villages and cities, where residents were unable to have cows of their own due to a lack of grazing land. Near the town, farmers could make some extra money on the side by having additional animals and selling the milk in town. The dairy farmers would fill barrels with milk in the morning and bring it to market on a wagon. Until the late 1800s, the milking of the cow was done by hand. In the United States, several large dairy operations existed in some northeastern states and in the west, that involved as many as several hundred cows, but an individual milker could not be expected to milk more than a dozen cows a day. Smaller operations predominated. Milking took place indoors in a barn with the cattle tied by the neck with ropes or held in place by stanchions. Feeding could occur simultaneously with milking in the barn, although most dairy cattle were pastured during the day between milkings. Such examples of this method of dairy farming are difficult to locate, but some are preserved as a historic site for a glimpse into the days gone by. One such instance that is open for this is at Point Reyes National Seashore.

Vacuum Bucket Milking

The first milking machines were an extension of the traditional milking pail. The early milker device fit on top of a regular milk pail and sat on the floor under the cow. Following each cow being milked, the bucket would be dumped into a holding tank. This developed into the Surge hanging milker. Prior to milking a cow, a large wide leather strap called a surcingle was put around the cow, across the cow's lower back. The milker device and collection tank hung underneath the cow from the strap. This innovation allowed the cow to move around naturally during the milking process rather than having to stand perfectly still over a bucket on the floor.

With the availability of electric power and suction milking machines, the production levels that were possible in stanchion barns increased but the scale of the operations continued to be limited by the labour intensive nature of the milking process. Attaching and removing milking machines involved repeated heavy lifting of the machinery and its contents several times per cow and the pouring of the milk into milk cans. As a result, it was rare to find single-farmer operations of more than 50 head of cattle.

Step-Saver Milk Transport

As herd size began to increase, the bucket milker system became laborious. A vacuum milk-transport system known as the Step-Saver was developed to transport milk to the storage tank. The system used a long vacuum hose coiled around a receiver cart, and connected to a vacuum-breaker device in the milk house, allowing farmers to milk many cows without the necessity of walking increasingly longer distances carrying heavy buckets of milk.

Milking Pipeline

The next innovation in automatic milking was the milk pipeline. This uses a permanent milk-return pipe and a second vacuum pipe that encircles the barn or milking parlour above the rows of cows, with quick-seal entry ports above each cow. By eliminating the need for the milk container, the milking device shrank in size and weight to the point where it could hang under the cow, held up only by the sucking force of the milker nipples on the cow's udder. The milk is pulled up into the milk-return pipe by the vacuum system, and then flows by gravity to the milk house vacuum-breaker that puts the milk in the storage tank. The pipeline system greatly reduced the physical labour of milking since the farmer no longer needed to carry around huge heavy buckets of milk from each cow.

The pipeline allowed barn length to keep increasing and expanding, but after a point farmers started to milk the cows in large groups, filling the barn with one-half to one-third of the herd, milking the animals, and then emptying and refilling the barn. As herd sizes continued to increase, this evolved into the more efficient milking parlour.

Milking Parlours

Innovation in milking focused on mechanising the milking parlour to maximise throughput of cows per operator which streamlined the milking process to permit cows to be milked as if on an assembly line, and to reduce physical stresses on the farmer by putting the cows on a platform slightly above the person milking the cows to eliminate having to constantly bend over. Many older and smaller farms still have tie-stall or stanchion barns, but worldwide a majority of commercial farms have parlours.

The milking parlour allowed a concentration of money into a small area, so that more technical monitoring and measuring equipment could be devoted to each milking station in the parlour. Rather than simply milking into a common pipeline for example, the parlour can be equipped with fixed measurement systems that monitor milk volume and record milking statistics for each animal. Tags on the animals allow the parlour system to automatically identify each animal as it enters the parlour.

Recessed Parlours

More modern farms use recessed parlours, where the milker stands in a recess such that his arms are at the level of the cow's udder. Recessed parlours can be herringbone, where the cows stand in two angled rows either side of the recess and the milker accesses the udder from the side, parallel, where the cows stand side-by-side and the milker accesses the udder from the rear or, more recently, rotary (or carousel), where the cows are on a raised circular platform, facing the centre of the circle, and the platform rotates while the milker stands in one place and accesses the udder from the rear. There are many other styles of milking parlours which are less common.

Herringbone and Parallel Parlours

In herringbone and parallel parlours, the milker generally milks one row at a time. The milker will move a row of cows from the holding yard into the milking parlour, and milk each cow in that row. Once all or most of the milking machines have been removed from the milked row, the milker releases the cows to their feed. A new group of cows is then loaded into the now vacant side and the process repeats until all cows are milked. Depending on the size of the milking parlour, which normally is the bottleneck, these rows of cows can range from four to sixty at a time.

Rotary Parlours

In rotary parlours, the cows are loaded one at a time onto the platform as it slowly rotates. The milker stands near the entry to the parlour and puts the cups on the cows as they move past. By the time the platform has completed almost a full rotation, another milker or a machine removes the cups and the cow steps backwards off the platform and then walks to its feed.

Automatic Milker take-off

It can be harmful to an animal for it to be over-milked past the point where the udder has stopped releasing milk. Consequently the milking process involves not just applying the milker, but also monitoring the process to determine when the animal has been milked out and the milker should be removed. While parlour operations allowed a farmer to milk many more animals much more quickly, it also increased the number of animals to be monitored simultaneously by the farmer. The automatic take-off system was developed to remove the milker from the cow when the milk flow reaches a preset level, relieving the farmer of the duties of carefully watching over 20 or more animals being milked at the same time.

Fully Automated Robotic Milking

In the 1980s and 1990s, robotic milking systems were developed and introduced (principally in the EU). Thousands of these systems are now in routine operation. In these systems the cow has a high degree of autonomy to choose her time of milking within pre-defined windows. These systems are generally limited to intensively managed systems although research continues to match them to the requirements of grazing cattle and to develop sensors to detect animal health and fertility automatically.

History of Milk Preservation Methods

Cool temperature has been the main method by which milk freshness has been extended. When windmills and well pumps were invented, one of its first uses on the farm besides providing water for animals was for cooling milk, to extend the storage life before being transported to the town market. The naturally cold underground water would be continuously pumped into a tub or other containers of milk set in the tub to cool after milking. This method of milk cooling was extremely popular before the arrival of electricity and refrigeration.

Refrigeration

When refrigeration first arrived (the 19th century) the equipment was initially used to cool cans of milk, which were filled by hand milking. These cans were placed into a cooled water bath to remove heat and keep them cool until they were able to be transported to a collection facility. As more automated methods were developed for harvesting milk, hand milking was replaced and, as a result, the milk can was replaced by a bulk milk cooler. 'Ice banks' were the first type of bulk milk cooler. This was a double wall vessel with evaporator coils and water located between the walls at the bottom and sides of the tank.

A small refrigeration compressor was used to remove heat from the evaporator coils. Ice eventually builds up around the coils, until it reaches a thickness of about three inches surrounding each pipe, and the cooling system shuts off. When the milking operation starts, only the milk agitator and the water circulation pump, which flows water across the ice and the steel walls of the tank, are needed to reduce the incoming milk to a temperature below 40 degrees.

This cooling method worked well for smaller dairies, however was fairly inefficient and was unable to meet the increasingly higher cooling demand of larger milking parlours. In the mid 1950's direct expansion refrigeration was first applied directly to the bulk milk cooler. This type of cooling utilizes an evaporator built directly into the inner wall of the storage tank to remove heat from the milk. Direct expansion is able to cool milk at a much faster rate than early ice bank type coolers and is still the primary method for bulk tank cooling today on small to medium sized operations.

Another device which has contributed significantly to milk quality is the plate heat exchanger (PHE). This device utilizes a number of specially designed stainless steel plates with small spaces between them. Milk is passed between every other set of plates with water being passed between the balance of the plates to remove heat from the milk.

This method of cooling can remove large amounts of heat from the milk in a very short time, thus drastically slowing bacteria growth and thereby improving milk quality. Ground water is the most common source of cooling medium for this device. Dairy cows consume approximately 3 gallons of water for every gallon of milk production and prefer to drink slightly warm water as opposed to cold ground water. For this reason, PHE's can result in drastically improved milk quality, reduced operating costs for the dairymen by reducing the refrigeration load on his bulk milk cooler, and increased milk production by supplying the cows with a source of fresh warm water.

Plate heat exchangers have also evolved as a result of the increase of dairy farm herd sizes in the US. As a dairyman increases the size of his herd, he must also increase the capacity of his milking parlour in order to harvest the additional milk. This increase in parlour sizes has resulted in tremendous increases in milk throughput and cooling demand. Today's larger farms produce milk at a rate which direct expansion refrigeration systems on bulk milk coolers cannot cool in a timely manner. PHE's are typically utilized in this instance to rapidly cool the milk to the desired temperature (or close to it) before it reaches the bulk milk tank. Typically, ground water is still utilized to provide some initial cooling to bring the milk to between 55 and 70 degrees F. A second (and sometimes third) section of the PHE is added to remove the remaining heat with a mixture of chilled pure water and propylene glycol. These chiller systems can be made to incorporate large evaporator surface areas and high chilled water flow rates to cool high flow rates of milk.

Milking Operation

Milking machines are held in place automatically by a vacuum system that draws the ambient air pressure down from 15 to 21 pounds per square inch (100 to 140 kPa) of vacuum. The vacuum is also used to lift milk vertically through small diameter hoses, into the receiving can. A milk lift pump draws the milk from the receiving can through large diameter stainless steel piping, through the plate cooler, then into a refrigerated bulk tank.

Milk is extracted from the cow's udder by flexible rubber sheaths known as liners or inflations that are surrounded by a rigid air chamber. A pulsating flow of ambient air and vacuum is applied to the inflation's air chamber during the milking process. When ambient air is allowed to enter the chamber, the vacuum inside the inflation causes the inflation to collapse around the cow's teat, squeezing the milk out of teat in a similar fashion as a baby calf's mouth massaging the teat. When the vacuum is reapplied in the chamber the flexible rubber inflation relaxes and opens up, preparing for the next squeezing cycle.

It takes the average cow three to five minutes to give her milk. Some cows are faster or slower. Slow-milking cows may take up to fifteen minutes to let down all their milk. Milking speed is only minorly related to the quantity of milk the cow produces — milking speed is a separate factor from milk quantity; milk quantity is not determinative of milking speed. Because most milkers milk cattle in groups, the milker can only process a group of cows at the speed of the slowest-milking cow. For this reason, many farmers will cull slow-milking cows.

The extracted milk passes through a strainer and plate heat exchangers before entering the tank, where it can be stored safely for a few days at approximately 42 °F (6 °C). At pre-arranged times, a milk truck arrives and pumps the milk from the tank for transport to a dairy factory where it will be pasteurized and processed into many products.

Animal Waste from Large Dairies

As measured in phosphorus, the waste output of 5,000 cows roughly equals a municipality of 70,000 people. In the U.S., dairy operations with more than 1,000 cows meet the EPA definition of a CAFO (Concentrated Animal Feeding Operation), and are subject to EPA regulations. For example, in the San Joaquin Valley of California a number of dairies have been established on a very large scale.

Each dairy consists of several modern milking parlour set-ups operated as a single enterprise. Each milking parlour is surrounded by a set of 3 or 4 loafing barns housing 1,500 or 2,000 cattle. Some of the larger dairies have planned 10 or more series of loafing barns and milking parlours in this arrangement, so that the total operation may include as many as 15,000 or 20,000 cows. The milking process for these dairies is similar to a smaller dairy with a single milking parlour but repeated several times.

The size and concentration of cattle creates major environmental issues associated with manure handling and disposal, which requires substantial areas of cropland (a ratio of 5 or 6 cows to the acre, or several thousand acres for dairies of this size) for manure spreading and dispersion, or several-acre methane digesters. Air pollution from methane gas associated with manure management also is a major concern. As a result, proposals to develop dairies of this size can be controversial and provoke substantial opposition from environmentalists including the Sierra Club and local activists. The potential impact of large dairies was demonstrated when a massive manure spill occurred on a 5,000-cow dairy in Upstate New York, contaminating a 20-mile (32 km) stretch of the Black River, and killing 375,000 fish. On Aug. 10, 2005, a manure storage lagoon collapsed releasing several million gallons of manure into the Black River. Subsequently the New York Department of Environmental Conservation mandated a settlement package of $2.2 million against the dairy.

Use of Hormones

It is possible to maintain higher milk production by injecting cows with growth hormones known as recombinant BST or rBGH, but this is controversial due to its effects on animal and possibly human health. The European Union, Japan, Australia, New Zealand and Canada have banned its use due to these concerns. However, no such prohibition exists in the US, where approximately 17.2% of dairy cows are treated in this way. The U.S. Food and Drug Administration states that no "significant difference" has been found between milk from treated and non-treated cows but based on consumer concerns several milk purchasers and resellers have elected not to purchase milk produced with rBST.

Management of the Herd

Modern dairy farmers use milking machines and sophisticated plumbing systems to harvest and store the milk from the cows, which are usually milked two or three times daily. During the warm months, in the northern hemisphere, cows may be allowed to graze in their pastures, both day and night, and are brought into the barn only to be milked. Many barns also incorporate tunnel ventilation into the architecture of the barn structure. This ventilation system is highly efficient and involves opening both ends of the structure allowing cool air to blow through the building. Farmers with this type of structure keep cows inside during the summer months to prevent sunburn and damage to udders.

During the winter months, especially in northern climates, the cows may spend the majority of their time inside the barn, which is warmed by their collective body heat. Even in winter, the heat produced by the cattle requires the barns to be ventilated for cooling purposes. Many modern facilities, and particularly those in tropical areas, keep all animals inside at all times to facilitate herd management. Housing the cow can be either loose housed or stalls (called cow cubicles in UK). There is little research available on dimensions required for cow stalls, and much housing can be out of date, however increasingly companies are making farmers aware of the benefits, in terms of animal welfare, health and milk production.

In the southern hemisphere cows spend most of their lives outside on pasture, although they may receive supplementation during periods of low pasture availabliity. The production of milk requires that the cow be in lactation, which is a result of the cow having given birth to a calf. The cycle of insemination, pregnancy, parturition, and lactation, followed by a "dry" period of a few weeks before calving which allows udder tissue to regenerate. Dairy operations therefore include both the production of milk and the production of calves. Bull calves are either castrated and raised as steers for beef production or veal.

Health and well-being

Common ailments affecting dairy cows include infectious disease (e.g. mastitis, endometritis and digital dermatitis), metabolic disease (e.g. milk fever and ketosis) and injuries caused by their environment (e.g. hoof and hock lesions). Lameness is commonly considered one of the most significant animal welfare issues for dairy cattle. It can be caused by a number of sources, including infections of the hoof tissue (e.g. fungal infections that cause dermatitis) and physical damage causing bruising or lesions (e.g. ulcers or hemorrhage of the hoof). While housing and management features common in modern conventional dairy farms (such as concrete barn floors, limited access to pasture and suboptimal bed-stall design) have been identified as contributing risk factors, small farms in developing countries can also demonstrate high rates.

Market

India is the largest producer of dairy products in the world. There is a great deal of variation in the pattern of dairy production worldwide. Many countries which are large producers, consume this internally, while others — in particular New Zealand — export a large percentage of their production. Internal consumption is often in the form of liquid milk, while the bulk of international trade is in processed dairy products such as milk powder. The world's largest exporter of dairy products is New Zealand, and dairy products are the largest export earner for the country. Fonterra is the fifth-largest dairy company in the world and New Zealand's largest company by turnover,

European Union

The European Union is the largest milk producer in the world, with 143.7 million tonnes in 2003. This data, encompassing the present 25 member countries, can be further broken down into the production of the original 15 member countries, with 122 million tonnes, and the new 10 mainly former Eastern European countries with 21.7 million tonnes.

United States

In the United States, the top four dairy states are, in order by total milk production; California, Wisconsin, New York, and Idaho. Dairy farming is also an important industry in Florida, Minnesota, Ohio and Vermont. There are 65,000 dairy farms in the United States.

Pennsylvania however, is the state with the heaviest dependence on dairy farming — there it is the number one industry. Pennsylvania is home to 8,500 farms and 555,000 dairy cows. Milk produced in Pennsylvania yields about US$1.5 billion in farm revenue every year, and is sold to various states up and down the east coast.

Milk prices collapsed in 2009. Senator Bernie Sanders accused Dean Foods of controlling 40% of the country's milk market. He has requested the United States Department of Justice to pursue 'an anti-trust investigation. Dean Foods says it buys 15% of the country's raw milk.

Competition

Most milk-consuming countries have a local dairy farming industry, and most producing countries maintain significant subsidies and trade barriers to protect domestic producers from foreign competition. In large countries, dairy farming tends to be geographically clustered in regions with abundant natural water supplies (both for feed crops and for cattle) and relatively inexpensive land (even under the most generous subsidy regimes, dairy farms have poor return on capital). New Zealand, the fourth largest dairy producing country, does not apply any subsidies to dairy production.

The milking of cows was traditionally a labour-intensive operation and still is in less developed countries. Small farms need several people to milk and care for only a few dozen cows, though for many farms these employees have traditionally been the children of the farm family, giving rise to the term "family farm".

Advances in technology have mostly led to the radical redefinition of "family farms" in industrialized countries such as the United States.

With farms of hundreds of cows producing large volumes of milk, the larger and more efficient dairy farms are more able to weather severe changes in milk price and operate profitably, while "traditional" very small farms generally do not have the equity or cash flow to do so.

The common public perception of large corporate farms supplanting smaller ones is generally a misconception, as many small family farms expand to take advantage of economies of scale, and incorporate the business to limit the legal liabilities of the owners and simplify such things as tax management. Before large scale mechanization arrived in the 1950s, keeping a dozen milk cows for the sale of milk was profitable. Now most dairies must have more than one hundred cows being milked at a time in order to be profitable, with other cows and heifers waiting to be "freshened" to join the milking herd. In New Zealand the average herd size, depending on the region, is about 350 cows.

Herd size in the US varies between 1,200 on the West Coast and Southwest, where large farms are commonplace, to roughly 50 in the Northeast, where land-base is a significant limiting factor to herd size. The average herd size in the U.S. is about one hundred cows per farm.

Currently, concerns regarding monopolies created by Dean Foods, Kraft, and other major buyers of bulk dairy products on the Chicago Mercantile Exchange have been raised, as American dairy farms have suffered extreme price depression and chaotic fluctuations while processors and retailers report record profits. Many theorize that unregulated imports of milk protein concentrate used by processors to boost cheese yield has artificially and unfairly influenced the markets in an effort to force consolidation and vertical integration in what has historically been a highly diversified industry.

Industrial Processing

Dairy plants process the raw milk they receive from farmers so as to extend its marketable life. Two main types of processes are employed: heat treatment to ensure the safety of milk for human consumption and to lengthen its shelf-life, and dehydrating dairy products such as butter, hard cheese and milk powders so that they can be stored.

Milk

Milk is an opaque white liquid produced by the mammary glands of mammals. It provides the primary source of nutrition for young mammals before they are able to digest other types of food. The early lactation milk is known as colostrum, and carries the mother's antibodies to the baby. It can reduce the risk of many diseases in the baby. The exact components of raw milk varies by species, but it contains significant amounts of saturated fat, protein and calcium as well as vitamin C. Cow's milk has a pH ranging from 6.4 to 6.8, making it slightly acidic.

Types of Consumption

There are two distinct types of milk consumption: a natural source of nutrition for all infant mammals, and a food product for humans of all ages derived from other animals.

Nutrition for Infant Mammals

In almost all mammals, milk is fed to infants through breastfeeding, either directly or by expressing the milk to be stored and consumed later. Some cultures, historically or currently, continue to use breast milk to feed their children until they are 7 years old.

In a short time, feeding infants fresh goat's milk in lieu of breast milk has become a common practice in Western culture. There are known dangers in this practice, including risk of developing electrolyte imbalances, metabolic acidosis, megaloblastic anemia and a host of allergic reactions.

Food Product for Humans

In many cultures of the world, especially the Western world, humans continue to consume milk beyond infancy, using the milk of other animals (especially cattle, goats and sheep) as a food product. For millennia, cow's milk has been processed into dairy products such as cream, butter, yogurt, kefir, ice cream, and especially the more durable and easily transportable product, cheese. Modern industrial processes produce casein, whey protein, lactose, condensed milk, powdered milk, and many other food-additive and industrial products.

Humans are an exception in the natural world for consuming milk past infancy, despite the fact that more than 75% of adult humans show some degree (some as little as 5%) of lactose intolerance, a characteristic that is more prevalent among individuals of African or Asian descent. The sugar lactose is found only in milk, forsythia flowers, and a few tropical shrubs.

The enzyme needed to digest lactose, lactase, reaches its highest levels in the small intestines after birth and then begins a slow decline unless milk is consumed regularly. On the other hand, those groups that do continue to tolerate milk often have exercised great creativity in using the milk of domesticated ungulates, not only of cattle, but also sheep, goats, yaks, water buffalo, horses, and camels. The largest producer and consumer of cattle and buffalo milk in the world is India.

Terminology

The term milk is also used for whitish non-animal substitutes such as soy milk, rice milk, almond milk, and coconut milk. Even the regurgitated substance secreted by glands in the mucosa of their upper digestive tract which pigeons feed their young is called crop milk though it bears little resemblance to mammalian milk.

Grade A vs Grade B milk

In the United States, there are two grades of milk, with Grade A primarily used for direct sales and consumption in stores, and Grade B used for indirect consumption, such as in cheese making or other processing. The differences between the two grades are defined in the Wisconsin administrative code for Agriculture, Trade, and Consumer Protection. Grade B generally refers to milk that is cooled in milk cans, which are immersed in a bath of cold flowing water, typically drawn up from an underground water well rather than using mechanical refrigeration;

- Grade A farms are inspected every 6 months, while Grade B farms are inspected every 2 years {WI-ATCP 60.24.2}
- Both types of farms are required to have two cleaning vats in the milk-house for washing and rinsing of equipment {WI-ATCP 60.07.2(g)}. A farm must also have an additional separate sink and faucet provided for handwashing {WI-ATCP 60.07.2(h)}, unless the bulk tank was installed before Jan 1, 1979 or the farm uses milk cans.
- Grade A milk stored in a bulk tank is cooled to 45 degrees F within 2 hours of milking. Grade A milk in a may only rise to 50 F if milk from additional milking sessions are added to the tank (potentially requiring a plate cooler to reduce the temperature of a large volume influx quickly enough) and must be cooled back to 45 F within two hours. {WI-ATCP 60.2.4(b)}
- Grade B milk in milk cans is cooled to 50 degrees F within 2 hours of milking. Grade B farms can not mix milk into cans from previous milking. {WI-ATCP 60.2.4(c)}
- The somatic cell count (SSC) of Grade A or B cow or sheep milk may not exceed 750,000 cells per mL, and the SCC of Grade A or B goat milk may not exceed 1,000,000 cells per mL. {WI-ATCP 60.15.4}
- The bacterial plate or loop count of Grade A milk may not exceed 100,000 per mL, while Grade B milk may not exceed 300,000 per mL. {WI-ATCP 60.15.2}

- A bacterial plate count test is required at least once a month. {WI-ATCP 60.18.3} If the bacterial count exceeds 100,000 per mL for Grade A or 300,000 per mL for grade B in 3 out of 5 tests, the license to sell milk is suspended. The license will be immediately revoked if the bacterial count ever exceeds 750,000 per mL. {WI-ATCP 60.18.6.

Evolution

Milk glands are highly specialized sweat glands. It has been suggested that the original function of lactation (milk production) was to keep eggs moist. Much of the argument is based on monotremes (egg-laying mammals):

History

Animal milk is first known to have been used as human food during the Secondary Products Revolution, around 5000BC. It is assumed that when animals such as cattle were first domesticated, it was only for purposes of meat. Cow's milk was first used as human food in the Middle East. Goats and sheep are ruminants: mammals adapted to survive on a diet of dry grass, a food source otherwise useless to humans, and one that is easily stockpiled.

The animals dairying proved to be a more efficient way of turning uncultivated grasslands into sustenance: the food value of an animal killed for meat can be matched by perhaps one year's worth of milk from the same animal, which will keep producing milk — in convenient daily portions — for years.

Milk byproducts found inside stone age pottery from Turkey indicate processed milk was consumed in 6500 BC some thousands of years before the ability for adult humans to digest unprocessed milk had evolved.

DNA evidence extracted from Neolithic skeletons indicates that a thousand years later in 5500 BC people in Northern Europe were like all other peoples of the time and were still lactose intolerant. Earthenware vessels found in England from a thousand years after this in 4500 BC contain milk byproducts indicating milk was used in some form although perhaps not drunk directly. Milk was first delivered in bottles on January 11, 1878. The day is now remembered as Milk Day and is celebrated annually. The town of Harvard, Illinois also celebrates milk in the summer with a festival known as "Milk Days". Theirs is a different tradition meant to celebrate dairy farmers in the "Milk Capital of the World."

Sources

In addition to cattle, the following livestock animals provide milk used by humans for dairy products:

- Camel
- Donkey
- Goat
- Horse
- Reindeer
- Sheep
- Water buffalo
- Yak.

According to the National Bison Association, American Bison (also called American buffalo) are not milked commercially. However, various sources report cows resulting from cross-breeding bison and domestic cattle are good milk producers, both during the European settlement of North America and during the development of commercial Beefalo in the 1970s and 1980s.

Human milk is not produced or distributed industrially or commercially; however, milk banks exist that allow for the collection of donated human milk and its redistribution to infants who may benefit from human milk for various reasons (premature neonates, babies with allergies or metabolic diseases, etc.).

All other female mammals do produce milk, but are rarely or never used to produce dairy products for human consumption.

Modern Production

In the Western world today, cow's milk is produced on an industrial scale, and is by far the most commonly consumed form of milk. Commercial dairy farming using automated milking equipment produces the vast majority of milk in developed countries.

Dairy cattle such as the Holstein have been specially bred for increased milk production. 90% of the dairy cows in the United States and 85% in Great Britain are Holsteins. Other dairy cows in the United States include Ayrshire, Brown Swiss, Guernsey, Jersey, and Milking Shorthorn (Dairy Shorthorn). The largest producers of dairy products and milk today are India followed by the United States, Germany, and Pakistan.

Increasing affluence in developing countries, as well as increased promotion of milk and milk products, has led to a rise in milk consumption in developing countries in recent years. In turn, the opportunities presented by these growing markets have attracted investment by multinational dairy firms.

Nevertheless, in many countries production remains on a small-scale and presents significant opportunities for diversification of income sources by small farmers. Local milk collection centers, where milk is collected and chilled prior to being transferred to urban dairies, are a good example of where farmers have been able to work on a cooperative basis, particularly in countries such as India.

Price

It was reported in 2007 that with increased worldwide prosperity and the competition of biofuel production for feedstocks, both the demand for and the price of milk had substantially increased world wide. Particularly notable was the rapid increase of consumption of milk in China and the rise of the price of milk in the United States above the government subsidized price.

Physical and Chemical Structure

Milk is an emulsion or colloid of butterfat globules within a water-based fluid. Each fat globule is surrounded by a membrane consisting of phospholipids and proteins; these emulsifiers keep the individual globules from joining together into noticeable grains of butterfat and also protect the globules from the fat-digesting activity of enzymes found in the fluid portion of the milk.

In unhomogenized cow's milk, the fat globules average about four micrometers across. The fat-soluble vitamins A, D, E, and K are found within the milkfat portion of the milk.

The largest structures in the fluid portion of the milk are casein protein micelles: aggregates of several thousand protein molecules, bonded with the help of nanometer-scale particles of calcium phosphate. Each micelle is roughly spherical and about a tenth of a micrometer across.

There are four different types of casein proteins, and collectively they make up around 80 percent of the protein in milk, by weight. Most of the casein proteins are bound into the micelles.

There are several competing theories regarding the precise structure of the micelles, but they share one important feature: the outermost layer consists of strands of one type of protein, k-Casein, reaching out from the body of the micelle into the surrounding fluid. These Kappa-casein molecules all have a negative electrical charge and therefore repel each other, keeping the micelles separated under normal conditions and in a stable colloidal suspension in the water-based surrounding fluid. Both the fat globules and the smaller casein micelles, which are just large enough to deflect light, contribute to the opaque white colour of milk. The fat globules contain some yellow-orange carotene, enough in some breeds (such as Guernsey and Jersey cattle) to impart a golden or "creamy" hue to a glass of milk. The riboflavin in the whey portion of milk has a greenish colour, which can sometimes be discerned in skimmed milk or whey products. Fat-free skimmed milk has only the casein micelles to scatter light, and they tend to scatter shorter-wavelength blue light more than they do red, giving skimmed milk a bluish tint.

Milk contains dozens of other types of proteins besides the caseins. They are more water-soluble than the caseins and do not form larger structures. Because these proteins remain suspended in the whey left behind when the caseins coagulate into curds, they are collectively known as whey proteins. Whey proteins make up around twenty percent of the protein in milk, by weight. Lactoglobulin is the most common whey protein by a large margin.

The carbohydrate lactose gives milk its sweet taste and contributes about 40% of whole cow's milk's calories. Lactose is a composite of two simple sugars, glucose and galactose. In nature, lactose is found only in milk and a small number of plants. Other components found in raw cow's milk are living white blood cells, Mammary-gland cells, various bacteria, and a large number of active enzymes.

Processing

In most Western countries, a centralized dairy facility processes milk and products obtained from milk (dairy products), such as cream, butter, and cheese. In the U.S., these dairies are usually local companies, while in the Southern Hemisphere facilities may be run by very large nationwide or trans-national corporations (such as Fonterra).

Pasteurization

Pasteurization is used to kill harmful microorganisms by heating the milk for a short time and then cooling it for storage and transportation. Pasteurized milk is still perishable and must be stored cold by both suppliers and consumers. Dairies print expiration dates on each container, after which stores will remove any unsold milk from their shelves.

A newer process, Ultra Pasteurization or ultra-high temperature treatment (UHT), heats the milk to a higher temperature for a shorter time. This extends its shelf life and allows the milk to be stored unrefrigerated because of the longer lasting sterilization effect.

Microfiltration

Microfiltration is a process that partially replaces pasteurization and produces milk with fewer microorganisms and longer shelf life without a change in the taste of the milk. In this process, cream is separated from the whey and is pasteurized in the usual way, but the whey is forced through ceramic microfilters that trap 99.9% of microorganisms in the milk (as compared to 95% killing of microorganisms in conventional pasteurization). The whey is then recombined with the pasteurized cream to reconstitute the original milk composition.

Creaming and Homogenization

Upon standing for 12 to 24 hours, fresh milk has a tendency to separate into a high-fat cream layer on top of a larger, low-fat milk layer. The cream is often sold as a separate product with its own uses; today the separation of the cream from the milk is usually accomplished rapidly in centrifugal cream separators. The fat globules rise to the top of a container of milk because fat is less dense than water. The smaller the globules, the more other molecular-level forces prevent this from happening. In fact, the cream rises in cow's milk much more quickly than a simple model would predict: rather than isolated globules, the fat in the milk tends to form into clusters containing about a million globules, held together by a number of minor whey proteins. These clusters rise faster than individual globules can. The fat globules in milk from goats, sheep, and water buffalo do not form clusters so readily and are smaller to begin with; cream is very slow to separate from these milks.

Milk is often homogenized, a treatment which prevents a cream layer from separating out of the milk. The milk is pumped at high pressures through very narrow tubes, breaking up the fat globules through turbulence and cavitation. A greater number of smaller particles possess more total surface area than a smaller number of larger ones, and the original fat globule membranes cannot completely cover them. Casein micelles are attracted to the newly-exposed fat surfaces; nearly one-third of the micelles in the milk end up participating in this new membrane structure. The casein weighs down the globules and interferes with the clustering that accelerated separation. The exposed fat globules are briefly vulnerable to certain enzymes present in milk, which could break down the fats and produce rancid flavors. To prevent this, the enzymes are inactivated by pasteurizing the milk immediately before or during homogenization. Homogenized milk tastes blander but feels creamier in the mouth than unhomogenized; it is whiter and more resistant to developing off flavors. Creamline, or cream-top, milk is unhomogenized; it may or may not have been pasteurized.

Milk which has undergone high-pressure homogenization, sometimes labelled as "ultra-homogenized," has a longer shelf life than milk which has undergone ordinary homogenization at lower pressures. Homogenized milk may be more digestible than unhomogenized milk.

Kurt A. Oster, M.D., who worked in the 1960s through the 1980s, suggested a link between homogenized milk and arterosclerosis, due to damage to plasmalogen as a result of the release of bovine xanthine oxidase (BXO) from the milk fat globular membrane (MFGM) during homogenization. However, Oster's hypothesis has been widely criticized and has not been generally accepted by the scientific community. No link has been found between arterosclerosis and milk consumption.

Nutrition and Health

The composition of milk differs widely between species. Factors such as the type of protein; the proportion of protein, fat, and sugar; the levels of various vitamins and minerals; and the size of the butterfat globules and the strength of the curd are among those than can vary. For example:

- Human milk contains, on average, 1.1% protein, 4.2% fat, 7.0% lactose (a sugar), and supplies 72 kcal of energy per 100 grams.
- Cow's milk contains, on average, 3.4% protein, 3.6% fat, and 4.6% lactose, 0.7% minerals and supplies 66 kcal of energy per 100 grams.

Donkey and horse milk have the lowest fat content, while the milk of seals and whales can contain more than 50% fat. High fat content is not unique to aquatic mammals, as guinea pig milk has an average fat content of 46%.

Processed milk began containing differing amounts of fat during the 1950s. 1 cup (250 ml) of 2%-fat milk contains 285 mg of calcium, which represents 22% to 29% of the daily recommended intake (DRI) of calcium for an adult. Depending on the age, milk contains 8 grams of protein, and a number of other nutrients (either naturally or through fortification) including:

- Biotin
- pantothenic acid

- Iodine
- Potassium
- Magnesium
- Selenium
- Thiamine
- Vitamin A
- Vitamin B12
- Riboflavin
- Vitamins D
- Vitamin K.

The amount of calcium from milk that is absorbed by the human body is disputed. Calcium from dairy products has a greater bioavailability than calcium from certain vegetables, such as spinach, that contain high levels of calcium-chelating agents, but a similar or lesser bioavailability than calcium from low-oxalate vegetables such as kale, broccoli, or other vegetables in the Brassica genus. The concentration of iodine in milk is usually higher than in the plasma of the mother, but may vary with her dietary iodine intake.

Medical Research

Studies show possible links between low-fat milk consumption and reduced risk of arterial hypertension, coronary heart disease, colorectal cancer and obesity. Overweight individuals who drink milk may benefit from decreased risk of insulin resistance and type 2 diabetes. One study has shown that for women desiring to have a child, those who consume full fat dairy products may actually slightly increase their fertility, while those consuming low fat dairy products may slightly reduce their fertility. Milk is a source of Conjugated linoleic acid.

Lactose Intolerance

Lactose, the disaccharide sugar component of all milk must be cleaved in the small intestine by the enzyme lactase in order for its constituents (galactose and glucose) to be absorbed. The production of this enzyme declines significantly after weaning in all mammals. Consequently, many humans become unable to properly digest lactose as they mature.

There is a great deal of variance, with some individuals reacting badly to even small amounts of lactose, some able to consume moderate quantities, and some able to consume large quantities of milk and other dairy products without problems. When an individual consumes milk without producing sufficient lactase, they may suffer diarrhea, intestinal gas, cramps and bloating, as the undigested lactose travels through the gastrointestinal tract and serves as nourishment for intestinal microflora who excrete gas, a process known as anaerobic respiration. It is estimated that 30 to 50 million Americans are lactose intolerant, including 75 percent of Native Americans and African-Americans, and 90 percent of Asian Americans. Lactose intolerance is less common among those descended from northern Europeans.

Lactose intolerance is a natural process and there is no reliable way to prevent or reverse it. Lactase is readily available in pill form, and many individuals can use it to briefly increase their tolerance for dairy products.

Controversy

Other studies suggest that milk consumption may increase the risk of suffering from certain health problems. Cow's milk allergy (CMA) is as an immunologically mediated adverse reaction to one or more cow's milk proteins. Rarely is it severe enough to cause death. Milk contains casein, a substance that breaks down in the human stomach to produce casomorphin, an opioid peptide. In the early 1990s it was hypothesized that casomorphin can cause or aggravate autism, and casein-free diets are widely promoted. Studies supporting these claims have had significant flaws, and the data are inadequate to guide autism treatment recommendations. Studies described in the book The China Study note a correlation between casein intake and the promotion of cancer cell growth when exposed to carcinogens. However other studies have shown whey protein offers a protective effect against colon cancer.

A study demonstrated that men who drink a large amount of milk and consume dairy products were at a slightly increased risk of developing Parkinson's disease; the effect for women was smaller. The reason behind this is not fully understood, and it also remains unclear why there is less of a risk for women.

Several sources suggest a correlation between high calcium intake (2000 mg per day, or twice the US recommended daily allowance, equivalent to six or more glasses of milk per day) and prostate cancer. A large study specifically implicates dairy, i.e., low-fat milk and other dairy to which vitamin A palmitate has been added.

A review published by the World Cancer Research Fund and the American Institute for Cancer Research states that at least eleven human population studies have linked excessive dairy product consumption and prostate cancer, however randomized clinical trial data with appropriate controls only exists for calcium, not dairy produce, where there was no correlation. Medical studies have also shown a possible link between milk consumption and the exacerbation of diseases such as Crohn's disease, Hirschsprung's disease–mimicking symptoms in babies with existing cow's milk allergies, and the aggravation of Behçet's disease.

Bovine Growth Hormone Supplementation

Since November 1993, with FDA approval, Monsanto has been selling recombinant bovine somatotropin (rbST), also called rBGH, to dairy farmers. Cows produce bovine growth hormone naturally, but some producers administer an additional recombinant version of BGH which is produced through a genetically-engineered E. coli because it increases milk production. Bovine growth horome also stimulates liver production of insulin-like growth factor 1 (IGF1). If rbST-treated cows produced milk with higher levels of IGF1 this would be of medical concern, because IGF1 stimulates cancer growth in humans.

Elevated levels of IGF1 in human blood has been linked to increased rates of breast, colon, and prostate cancer. Monsanto has stated that both of these compounds are harmless given the levels found in milk and the effects of pasteurization. However Monsanto's own tests, conducted in 1987, demonstrated that statistically significant growth stimulating effects were induced in organs of adult rats by feeding IGF-1 at low dose levels for only two weeks. "Drinking rBGH milk would thus be expected to significantly increase IGF-1 blood levels and consequently to increase risks of developing breast cancer and promoting its invasiveness."

The EU has recommended against Monsanto milk On June 9, 2006, the largest milk processor in the world and the two largest supermarkets in the United States—Dean Foods, Wal-Mart, and Kroger—announced that they are "on a nationwide search for rBGH-free milk." Milk from cows given rBST may be sold in the United States, and the FDA stated that no significant difference has been shown between milk derived from rBST-treated and that from non-rBST-treated cows. Milk that advertises that it comes from cows not treated with rBST is required to state this finding on its label.

Cows receiving rBGH supplements may more frequently contract an udder infection known as mastitis. Problems with mastitis have led to Canada, Australia, New Zealand, and Japan banning milk from rBST treated cows. Mastitis, among other diseases, may be responsible for the fact that levels of white blood cells in milk vary naturally.

Ethical Concerns

Vegans and some other vegetarians do not consume milk for a variety of reasons. They may object to features of dairy farming including the necessity of killing almost all the male offspring of dairy cows (either by disposal soon after birth, for veal production, or for beef), the routine separation of mother and calf soon after birth, other perceived inhumane treatment of dairy cattle, and culling of cows after their productive lives.

Varieties and Brands

Milk products are sold in a number of varieties based on types/degrees of:

- age (e.g., cheddar),
- additives (e.g., vitamins),
- coagulation (e.g., cottage cheese),
- farming method (e.g., organic, grass-fed).
- fat content (e.g., half and half),
- fermentation (e.g., buttermilk),
- flavouring (e.g., chocolate),
- homogenization (e.g., raw milk),
- mammal (e.g., cow, goat, sheep),
- packaging (e.g., bottle),
- sterilization (e.g., pasteurization),
- water content (e.g., dry milk).

Demeter certified milk is produced with biodynamic agriculture methods and is similar in standards to organic milk and biological milk, with a few special farm procedures added that are biodynamic-specific.

Additives and Flavouring

In areas where the cattle (and often the people) live indoors, commercially sold milk commonly has vitamin D added to it to make up for lack of exposure to UVB radiation. Reduced fat milks often have added vitamin A palmitate to compensate for the loss of the vitamin during fat removal; in the United States this results in reduced fat milks having a higher vitamin A content than whole milk.

To aid digestion in those with lactose intolerance, milk is available in some areas with added bacterial cultures such as Lactobacillus acidophilus ("acidophilus milk") and bifidobacteria ("a/B milk"). Another milk with Lactococcus lactis bacteria cultures ("cultured buttermilk") is often used in cooking to replace the traditional use of naturally soured milk, which has become rare due to the ubiquity of pasteurization which kills the naturally occurring lactococcus bacteria.

Milk often has flavouring added to it for better taste or as a means of improving sales. Chocolate milk has been sold for many years and has been followed more recently by strawberry milk and others. Some nutritionists have criticized flavoured milk for adding sugar, usually in the form of high fructose corn syrup, to the diets of children who are already commonly obese in the US.

Distribution

Because milk spoils so easily, it should, ideally, be distributed as quickly as possible. In many countries milk used to be delivered to households daily, but economic pressure has made milk delivery much less popular, and in many areas daily delivery is no longer available. People buy it chilled at grocery or convenience stores or similar retail outlets. Prior to the widespread use of plastics, milk was sold in wax-coated paper containers; prior to that milk was often distributed to consumers in glass bottles; and before glass bottles, in bulk that was ladled into the customer's container.

United Kingdom

In the UK, milk can be delivered daily by a milkman who travels his local milk round (route) using a milk float (often battery powered) during the early hours. Milk is delivered in 1 pint glass bottles with aluminium foil tops. Silver top denotes full cream unhomogenized; red top full cream homogenized; red/silver top semi-skimmed; blue/silver check top skimmed; and gold top channel island.

Empty bottles are rinsed before being left outside for the milkman to collect and take back to the dairy for washing and reuse. Currently many milkmen operate franchises as opposed to being employed by the dairy and payment is made at regular intervals, by leaving a cheque; by cash collection; or direct debit.

Although there was a steep decline in doorstep delivery sales throughout the 1990s, the service is still prominent, as dairies have diversified and the service is becoming more popular again. The doorstep delivery of milk is seen as part of the UK's heritage, and is used by people up and down the country.

Australia and New Zealand

In Australia and New Zealand, prior to "metrification", milk was generally distributed in 1 pint (568ml) glass bottles. In Australia there was a government funded "free milk for school children" program, and milk was distributed at morning recess in 1/3 pint bottles.

With the conversion to metric measures, the milk industry were concerned that the replacement of the pint bottles with 500ml bottles would result in a 13.6% drop in milk consumption. Hence, all pint bottles were recalled and replaced by 600ml bottles. With time, due to the steadily increasing cost of collecting, transporting, storing and cleaning glass bottles, they were replaced by cardboard cartons.

A number of designs were used, including a tetrahedron which could be close-packed without waste space, and could not be knocked over accidentally. (slogan: No more crying over spilt milk.)

However, the industry eventually settled on a design similar to that used in the United States. Milk is now availability in a variety of sizes in cardboard cartons (250ml, 375ml, 600ml, 1 litre and 1.5 litres) and plastic bottles (1 in NZ, 1.1 in Australia, 2 and 3 litres). A significant addition to the marketplace has been "long life" milk (UHT), generally available in 1 and 2 litre rectangular cardboard cartons. In urban and suburban areas where there is sufficient demand, home delivery is still available, though in suburban areas this is often 3 times per week rather than daily. Another significant and popular addition to the marketplace has been flavoured milks-for example, as mentioned above, Farmers Union Iced Coffee outsells Coca-Cola in South Australia.

India

In rural India milk is delivered daily by a local milkman carrying bulk quantities in a metal container, usually on a bicycle; and in other parts of metropolitan India, milk is usually bought or delivered in a plastic bags or cartons via shops or supermarkets.

United States

In the United States, glass milk bottles have been mostly replaced with milk cartons (tall paper boxes with a square cross-section and a peaked top that can be folded outward upon opening to form a spout) and plastic jugs.

Gallons of milk are almost always sold in jugs, while half-gallons and quarts may be found in both paper cartons and plastic jugs, and smaller sizes are almost always in cartons. Recently, milk has been sold in smaller resealable bottles made to fit in automobile cup holders. These individual serving sizes are also sold in flavoured varieties. The half-pint milk carton is the traditional unit as a component of school lunches.

Pakistan

In Pakistan, milk is supplied in jugs. Milk has been a staple food, especially among the pastoral tribes in this country.

UHT Milk

Milk preserved by the UHT process is sold in cartons often called a brick that lack the peak of the traditional milk carton. Milk preserved in this fashion does not need to be refrigerated before opening and has a longer shelf life than milk in ordinary packaging. It is more typically sold unrefrigerated on the shelves in Europe and Latin America than in the United States. In Australia it is generally sold unrefrigerated, though some supermarkets also keep small quantities refrigerated.

Glass

Glass milk containers are now rare. Most people purchase milk in bags, plastic bottles, or plastic-coated paper cartons. Ultraviolet (UV) light from fluorescent lighting can alter the flavour of milk, so many companies that once distributed milk in transparent or highly translucent containers are now using thicker materials that block the UV light. Many people feel that such "UV protected" milk tastes better.

Australia and New Zealand: Distributed in a variety of sizes, most commonly in aseptic cartons for up to 1.5 litres, and plastic screw-top bottles beyond that with the following volumes; 1.1L, 2L, and 3L. 1 litre milk bags are starting to appear in supermarkets, but have not yet proved popular. Most UHT-milk is packed in 1 or 2 litre paper containers with a sealed plastic spout.

Brazil :Used to be sold in cooled 1 litre bags, just like in South Africa. Nowadays the most common form is 1 litre aseptic cartons containing UHT skimmed, semi-skimmed or whole milk, although the plastic bags are still in use for pasteurized milk. Higher grades of pasteurized milk can be found in cartons or plastic bottles. Sizes other than 1 liter are rare.

Canada: 1.33 litre plastic bags (sold as 4 litres in 3 bags) are widely available in some areas (especially the Maritimes, Ontario and Quebec), although the 4 litre plastic jug has supplanted them in western Canada. Other common packaging sizes are 2 litre, 1 litre, 500 millilitre, and 250 millilitre cartons, as well as 4 litre, 1 litre, 250 ml aseptic cartons and 500 ml plastic jugs.

China: Sweetened milk is a drink popular with students of all ages and is often sold in small plastic bags complete with straw. Adults not wishing to drink at a banquet often drink milk served from cartons or milk tea.

Croatia, Bosnia and Herzegovina, Serbia, Montenegro: *UHT milk is sold in 500 ml and 1 L (sometimes also 200 ml) aseptic cartons. Non-UHT pasteurized milk is most commonly sold in 1 L and 1.5 L PET bottles, though in Serbia one can still find milk in plastic bags.*

Parts of Europe: Sizes of 500 millilitres, 1 litre (the most common), 1.5 litres, 2 litres and 3 litres are commonplace.

Finland: Commonly sold in 1l or 1.5l cartons, in some places also in 2dl and 5dl cartons.

Hong Kong: milk is sold in glass bottles (220 ml), cartons (236 ml and 1L), plastic jugs (2 litres) and aseptic cartons (250 ml).

India : Commonly sold in 500 ml plastic bags and in bottles in some parts like in west. It is still customary to serve the milk boiled, despite pasteurization. Milk is often buffalo milk. Flavoured milk is sold in most convenience stores in waxed cardboard containers. Convenience stores also sell many varieties of milk (such as flavoured and ultra-pasteurized) in different sizes, usually in aseptic cartons.

Indonesia: Usually sold in 1 litre cartons, but smaller, snack-sized cartons are available.

Israel : Non-UHT milk is most commonly sold in 1 litre waxed cardboard boxes and 1 litre plastic bags. It may also be found in 0.5L and 2L waxed cardboard boxes, 2L plastic jugs and 1L plastic bottles. UHT milk is available in 1 litre (and less commonly also in 0.25L) carton "bricks".

Japan : Commonly sold in 1 litre waxed paperboard cartons. In most city centers there is also home delivery of milk in glass jugs. As seen in China, sweetened and flavoured milk drinks are commonly seen in vending machines.

South Africa: Commonly sold in 1 litre bags. The bag is then placed in a plastic jug and the corner cut off before the milk is poured.

South Korea: sold in cartons (180ml, 200ml, 500ml 900ml, 1L, 1.8L, 2.3L), plastic jugs (1L and 1.8L), aseptic cartons (180ml and 200ml) and plastic bags (1L).

Sweden: Commonly sold in 0.3L, 1L or 1.5L cartons and sometimes as plastic or glass milk bottles..

Poland: UHT milk is mostly sold in aseptic cartons (500ml, 1L, 2L), and non-UHT in 1L plastic bags or plastic bottles. Milk, UHT is commonly boiled, despite being pasteurized.

Pakistan: Milk is supplied in 500 ml Plastic bags and carried in Jugs from rural to cities and sell

Turkey: Commonly sold in 500 ml or 1L cartons or special plastic bottles. UHT milk is more popular. Milkmen also serve in smaller towns and villages.

United Kingdom: Most stores stock imperial sizes: 1 pint (568 ml), 2 pints (1.136 L), 4 pints (2.273 L), 6 pints (3.408 L) or a combination including both metric and imperial sizes. Glass milk bottles delivered to the doorstep by the milkman are typically pint-sized and are returned empty by the householder for repeated reuse. Milk is sold at supermarkets in either aseptic cartons or HDPE bottles. Milk can still be legally sold by the imperial pint in the UK under EU regulations (a distinction only shared with beer and cider).

United States : Commonly sold in gallon (3.78 L), half-gallon (1.89 L) and quart (0.94 L) containers of natural-coloured HDPE resin, or, for sizes less than one gallon, cartons of waxed paperboard. Bottles made of opaque PET are also becoming commonplace for smaller, particularly metric, sizes such as one liter. The U.S. single-serving size is usually the half-pint (about 240 mL). Less frequently, dairies deliver milk directly to consumers, from coolers filled with glass bottles which are typically half-gallon sized and returned for reuse. Some convenience store chains in the United States (such as Kwik Trip in the Midwest) sell milk in half-gallon bags.

Uruguay: Commonly sold in 1 litre bags. The bag is then placed in a plastic jug and the corner cut off before the milk is poured.

Practically everywhere, condensed milk and evaporated milk is distributed in metal cans, 250 and 125 ml paper containers and 100 and 200 ml squeeze tubes, and powdered milk (skim and whole) is distributed in boxes or bags.

Spoilage and Fermented Milk Products

When raw milk is left standing for a while, it turns "sour". This is the result of fermentation, where lactic acid bacteria ferment the lactose inside the milk into lactic acid. Prolonged fermentation may render the milk unpleasant to consume.

This fermentation process is exploited by the introduction of bacterial cultures (e.g. Lactobacilli sp., Streptococcus sp., Leuconostoc sp., etc.) to produce a variety of fermented milk products. The reduced pH from lactic acid accumulation denatures proteins and caused the milk to undergo a variety of different transformations in appearance and texture, ranging from an aggregate to smooth consistency. Some of these products include sour cream, yoghurt, cheese, buttermilk, viili, kefir and kumis. Pasteurization of cow's milk initially destroys any potential pathogens and increases the shelf-life, but eventually results in spoilage that makes it unsuitable for consumption. This causes it to assume an unpleasant odour, and the milk is deemed non-consumable due to unpleasant taste and an increased risk of food poisoning. In raw milk, the presence of lactic acid-producing bacteria, under suitable conditions, ferments the lactose present to lactic acid. The increasing acidity in turn prevents the growth of other organisms, or slows their growth significantly. During pasteurization however, these lactic acid bacteria are mostly destroyed.

In order to prevent spoilage, milk can be kept refrigerated and stored between 1 and 4 degrees Celsius in bulk tanks. Most milk is pasteurized by heating briefly and then refrigerated to allow transport from factory farms to local markets. The spoilage of milk can be forestalled by using ultra-high temperature (UHT) treatment; milk so treated can be stored unrefrigerated for several months until opened. Condensed milk, made by removing most of the water, can be stored in cans for many years, unrefrigerated, as can evaporated milk. The most durable form of milk is milk powder, which is produced from milk by removing almost all water. The moisture content is usually less than five percent in both drum and spray dried milk powder.

Language and Culture

The importance of milk in human culture is attested to by the numerous expressions embedded in our languages, for example "the milk of human kindness". In ancient Greek mythology, the goddess Hera spilled her breast milk after refusing to feed Heracles, resulting in the Milky Way.

In African and Asian developing nations, butter is traditionally made from fermented milk rather than cream. It can take several hours of churning to produce workable butter grains from fermented milk.

Holy books have also mentioned milk; the Bible contains references to the 'Land of Milk and Honey'. In the Quran, there is a request to wonder on milk as follows: 'And surely in the livestock there is a lesson for you, We give you to drink of that which is in their bellies from the midst of digested food and blood, pure milk palatable for the drinkers.'(16-The Honeybee, 66). The Ramadhan fast is traditionally broken with a glass of milk and dates.

The verb, "to milk" something is often used in the vernacular of many English-speaking countries as a synonym for extortion or, in less loaded terms, taking advantage of a situation where one has another person at a disadvantage, as in 'milking the situation'. The word milk has had many slang meanings over time. In the early 17th century the word was used to mean semen, or vaginal secretions, or to masturbate oneself or someone else. In the 19th century, milk was used to describe a cheap alcoholic drink made from methylated spirits mixed with water. The word was also used to mean defraud, to be idle, to intercept telegrams addressed to someone else, and a weakling or 'milksop'. In the mid 1930s, the word was used in Australia meaning to siphon gas from a car.

Milk is sometimes referred to as moo juice in American English, while Cockney rhyming slang calls it Acker Bilk, Tom Silk, Lady in silk and Kilroy Silk. The name of the Russian Molokan religion in Russian is derived from Russian "Ìîëîêî " meaning "Milk" as they would drink milk on the Russian Orthodox days of fast.

Other Uses

Besides serving as a beverage or source of food, milk has been described as used by farmers and gardeners as an organic fungicide and foliage fertilizer. Diluted milk solutions have been demonstrated to provide an effective method of preventing powdery mildew on grape vines, while showing it is unlikely to harm the plant.

Cream and Butter

Today, milk is separated by large machines in bulk into cream and skim milk. The cream is processed to produce various consumer products, depending on its thickness, its suitability for culinary uses and consumer demand, which differs from place to place and country to country. Some cream is dried and powdered, some is condensed (by evaporation) mixed with varying amounts of sugar and canned. Most cream from New Zealand and Australian factories is made into butter. This is done by churning the cream until the fat globules coagulate and form a monolithic mass. This butter mass is washed and, sometimes, salted to improve keeping qualities. The residual buttermilk goes on to further processing. The butter is packaged (25 to 50 kg boxes) and chilled for storage and sale. At a later stage these packages are broken down into home-consumption sized packs. Butter sells for about US$3200 a tonne on the international market in 2007 (an unusual high).

Skimmed Milk

The product left after the cream is removed is called skim, or skimmed, milk. Reacting skim milk with rennet or with an acid makes casein curds from the milk solids in skim milk, with whey as a residual. To make a consumable liquid a portion of cream is returned to the skim milk to make low fat milk (semi-skimmed) for human consumption. By varying the amount of cream returned, producers can make a variety of low-fat milks to suit their local market. Other products, such as calcium, vitamin D, and flavouring, are also added to appeal to consumers.

Casein

Casein is the predominant phosphoprotein found in fresh milk. It has a very wide range of uses from being a filler for human foods, such as in ice cream, to the manufacture of products such as fabric, adhesives, and plastics. However, in the United States these assorted non-food uses have led to concerns over the import of substandard (non-food-grade) powders from other countries, such as China, that are then used to artificially bolster domestic cheese yield without the casein additive undergoing Food and Drug Administration inspection.

Cheese

Cheese is another product made from milk. Whole milk is reacted to form curds that can be compressed, processed and stored to form cheese. In countries where milk is legally allowed to be processed without pasteurisation a wide range of cheeses can be made using the bacteria naturally in the milk. In most other countries, the range of cheeses is smaller and the use of artificial cheese curing is greater. Whey is also the by-product of this process.

Cheese has historically been an important way of "storing" milk over the year, and carrying over its nutritional value between prosperous years and fallow ones. It is a food product that, with bread and beer, dates back to prehistory in Middle Eastern and European cultures, and like them is subject to innumerable variety and local specificity. Although nowhere near as big as the market for cow's milk cheese, a considerable amount of cheese is made commercially from other milks, especially goat and sheep.

Milk Powders

Milk is also processed by various drying processes into powders. Whole milk, skim milk, buttermilk, and whey products are dried into a powder form and used for human and animal consumption. The main difference between production of powders for human or for animal consumption is in the protection of the process and the product from contamination. Some people drink milk reconstituted from powdered milk, because milk is about 88% water and it is much cheaper to transport the dried product. Dried skim milk powder is worth about US$5300 a tonne (mid-2007 prices) on the international market.

Other Milk Products

Kumis is produced commercially in Central Asia. Although it is traditionally made from mare's milk, modern industrial variants may use cow's milk instead.

Transport of Milk

Historically, the milking and the processing took place in the same place: on a dairy farm. Later, cream was separated from the milk by machine, on the farm, and the cream was transported to a factory for buttermaking. The skim milk was fed to pigs. This allowed for the high cost of transport (taking the smallest volume high-value product), primitive trucks and the poor quality of roads. Only farms close to factories could afford to take whole milk, which was essential for cheese making in industrial quantities, to them. The development of refrigeration and better road transport, in the late 1950s, has meant that most farmers milk their cows and only temporarily store the milk in large refrigerated bulk tanks, whence it is later transported by truck to central processing facilities.

Milking Machines

Milking machines are used to harvest milk from cows when manual milking becomes inefficient or labour intensive. The milking unit is the portion of a milking machine for removing milk from an udder. It is made up of a claw, four teatcups, (Shells and rubber liners) long milk tube, long pulsation tube, and a pulsator. The claw is an assembly that connects the short pulse tubes and short milk tubes from the teatcups to the long pulse tube and long milk tube. (Cluster assembly) Claws are commonly made of stainless steel or plastic or both. Teatcups are composed of a rigid outer shell (stainless steel or plastic) that holds a soft inner liner or inflation. Transparent sections in the shell may allow viewing of liner collapse and milk flow. The annular space between the shell and liner is called the pulse chamber.

Milking machines work in a way that is different from hand milking or calf suckling. Continuous vacuum is applied inside the soft liner to massage milk from the teat by creating a pressure difference across the teat canal (or opening at the end of the teat). Vacuum also helps keep the machine attached to the cow. The vacuum applied to the teat causes congestion of teat tissues (accumulation of blood and other fluids). Atmospheric air is admitted into the pulsation chamber about once per second (the pulsation rate) to allow the liner to collapse around the end of teat and relieve congestion in the teat tissue. The ratio of the time that the liner is open (milking phase) and closed (rest phase) is called the pulsation ratio.

The four streams of milk from the teatcups are usually combined in the claw and transported to the milkline, or the collection bucket (usually sized to the output of one cow) in a single milk hose. Milk is then transported (manually in buckets) or with a combination of airflow and mechanical pump to a central storage vat or bulk tank. Milk is refrigerated on the farm in most countries either by passing through a heat-exchanger or in the bulk tank, or both.

In the photo above is a bucket milking system with the stainless steel bucket visible on the far side of the cow. The two rigid stainless steel teatcup shells applied to the front two quarters of the udder are visible. The top of the flexible liner is visible at the top of the shells as are the short milk tubes and short pulsation tubes extending from the bottom of the shells to the claw. The bottom of the claw is transparent to allow observation of milk flow. When milking is completed the vacuum to the milking unit is shut off and the teatcups are removed.

Milking machines keep the milk enclosed and safe from external contamination. The interior 'milk contact' surfaces of the machine are kept clean by a manual or automated washing procedures implemented after milking is completed. Milk contact surfaces must comply with regulations requiring food-grade materials (typically stainless steel and special plastics and rubber compounds) and are easily cleaned.

Most milking machines are powered by electricity but, in case of electrical failure, there can be an alternative means of motive power, often an internal combustion engine, for the vacuum and milk pumps. Milk cows cannot tolerate delays in scheduled milking without serious milk production reductions.

Chapter 10

Reproductive Methods in Low Input Animal Breeding

Generally we should declare that ethics cannot be described by exact parameters but it is the conclusion of some principals and long lasting social rules directing human behaviour.

It is a similar phenomenon in animal breeding and mostly in that type of breeding which has historical traditions and emotional relation to country people.

Since ethical issues have been getting rather "artificial" explanation by the media recently, people in the cities and unfortunately sometimes in the countryside many times do not have proper information on this field.

Nevertheless reproductive researchers also do not have opinion overcoming all others usually they have a modest and realistic behaviour in their work aiming to improve the farmers' results in their daily practice.

Ethical aspects of reproductive methods in low input breeding (LIB) need a rather complex approach determined by:

1. Human factors – rural local and regional healthy food supply, rural development i.e. reducing unemployment, maintaining agriculture traditions, support rural tourism and environment protection.
2. Animal factors – gene conservation by all possible and necessary methods as well as animal welfare in the balanced frame of traditions and actual demands according to the forthcoming challenges of the rapidly changing world.

Easy to see that even by the applied reproductive management we cannot respond all criteria Sabove thus we should select the priorities with regard on main purpose of the farm in case. Before discussing the different techniques it should be declared that reproductive methods even in LIB should not miss the well trained experts having relevant knowledge and experience. Animal health conditions, individual registration of animals, exact documentation of mating or artificial insemination (AI) are required at the same manner as in intensive farming. It means that professionals involved in LIB should be familiar not only with the traditional methods but with the newest results of innovation, as well.

We try to give over some of our experiences regarding on reproductive techniques in LIB which has a special importance also in Hungary and we were conducting international research projects connected to it. Animal breeds in LIB are mostly indigenous and sometimes endangered ones adapted to the climate of Carpathian Basin very well during recent centuries. These animals are incorporating our traditions, representing our agricultural national value and capable for playing a key role in rural development, rural tourism and first of all unique processed products can be produced of their carcass i.e. Hungarian winter salami, sausage, bacon etc.

Mangalica Pig

Mangalica pigs, more precisely Blond, Red and Swallow Belly Mangalica were always bred in small and large scale farms in the past and it is the same in the present. Apart from their own production large companies integrated small farmers' activity and organized the breeding and trading. The Mangalica Breeding Association is giving guidance for the members and supply boars or semen for them.

Reproductive work is done by natural mating and AI both in small and large farms. In some blood strains the low number of animals evidently needs natural mating for in vivo gene preservation. In some production units AI is a daily practice.

We should underline the necessity of AI in any production units. It is not so much costly as keeping minimally 5 times more boars but highly dedicated work is indispensable. Generally the AI is significantly different and probably more complicated in native breeds than in their modern counterparts. It is true in terms of male as well as female reproductive physiology. Semen deep freezing is a connecting wing of LIB when involved farmers and companies are cooperating with research units and they develop together the most important tool for in vitro gene conservation. Although, boar semen cryopreservation is not clearly solved, our group has some really promising results (average 50% post-thaw motility).

National parks have an emphasized duty to demonstrate our agriculture traditions and they should keep indigenous domestic animal breeds e.g. Mangalica pigs in pure bred population among old, typical LIB circumstances. It can be concluded that LIB in Hungarian Mangalica sector found its proper position and small farmers can take part of local and regional food supply by high value processed pork products. (Unfortunately recently upcoming high feed prices are very harmful for Mangalica breeders, too.)

Several successful attempts are running in Europe using also commercial breeds in organic farming for producing "BIO" labelled meat and processed products.

Racka Sheep

Racka sheep has history of several hundred years. The Hungarian and other nomad tribes had similar type of sheep already on the Middle Asian steppes and we can find nearly identical breeds there even now. It has a diminished population in Hungary with two colour types i.e. the White and Black Racka. National parks and enthusiastic sheep breeders keep, also some village hotels keep them in small units. Although some attempts have been done to establish a new market role for it lower meat yield % (in EUROPE classification) undermined the efforts till now.

Definitely an other evaluation system would be desirable for the ancient breeds. Anyway, Racka is excellently suitable for LIB in continental climate of the Carpathian Basin thus it could contribute rural development programs in remote areas.

Actually eminent purpose of reproductive management declared by the Hungarian Sheep Breeders' Association is the preservation of this breed. Almost everywhere ewes are mated naturally by selected rams.

LIB has a special significance in modern animal breeding. Both in rural tourism and rural development it can find its proper place but farmers involved should clearly know their role inthe sector. If they choose small scale farming, they must use the relevant techniques e.g. reproductive methods and if they decide to take part of larger production amount they should adapt modern and more effective system. Innovation is always necessary either by taking part in it or by collecting the available achievements. In our consideration reproductive methods in LIB does not mean closed eyes and ears but to be opened for the new results enabling the farmers to improve their dedicated work.

Genetic Engineering Applications in Animal Breeding

Genetic engineering is the name of a group of techniques used for direct genetic modification of organisms or population of organisms using recombination of DNA. These procedures are of use to identify, replicate, modify and transfer the genetic material of cells, tissues or complete organisms. Most techniques are related to the direct manipulation of DNA oriented to the expression of particular genes. In a broader sense, genetic engineering involves the incorporation of DNA markers for selection (marker-assisted selection, MAS), to increase the efficiency of the so called 'traditional' methods of breeding based on phenotypic information. The most accepted purpose of genetic engineering is focused on the direct manipulation of DNA sequences These techniques involve the capacity to isolate, cut and transfer specific DNA pieces, corresponding to specific genes.

The mammalian genome has a larger size and has a more complex organization than in viruses, bacteria and plants. Consequently, genetic modification of animals, using molecular genetics and recombinant DNA technology is more difficult and costly than in simpler organisms. In mammals, techniques for reproductive manipulation of gametes and embryos such as obtaining of a complete new organism from adult differentiated cells (cloning), and procedures for artificial reproduction such as in vitro fertilization, embryo transfer and artificial insemination, are frequently an important part of these processes.

Current research in genetic engineering of animals is oriented toward a variety of possible medical, pharmaceutical and agricultural applications. Also, there is an interest to increase basic knowledge about mammalian genetics and physiology, including complex traits controlled by many genes such as many human and animal diseases. The interest in genetic engineering of mammalian cells is based in the idea of, for example, use gene therapy to cure genetic diseases such as cystic fibrosis by replacing the damaged copies of the gene by normal ones in foetuses or infants (gene therapy).

Genetically engineered animals such as the 'knockout mouse', in which one specific gene is 'turned off', are used to model genetic diseases in humans and to discover the function of specific sites of the genome (Majzoub and Muglia, 1996). Genetically modified animals such as pigs will probably be used to produce organs for transplant to humans (xenotransplantation).

Other applications include production of specific therapeutic human proteins such as insulin in the mammary gland of genetically modified milking animals like goats (transgenic animals, bioreactors). These techniques may be used to increase disease resistance and productivity in agriculturally important animals by increasing the frequency of the desired alleles in the populations used in food production. This can be accomplished by transferring alleles or allele combinations, over expressing or eliminating the expression of particular genes (use of genetic engineering in animal breeding). In addition, these techniques open the possibility of using artificially modified genes to increase the biological efficiency of proteins (Kinghorn, 2003).

The objective of this paper is to review some advances on genetic engineering applications in animal breeding, including a description of the methods, some applications and ethical issues. Here I made emphasis in both the search and use of genomic information for selecting animals and to transfer and use their genes in commercial populations via marker-assisted selection (MAS) or transgenesis. This review focuses mainly in the methodology to apply genetic engineering directly to animals for genetic improvement.

Several important biotechnological applications such as the production of recombinant proteins in bioreactors (Houdebine, 2002), disease diagnostic (McKeever and Rege, 1999), feedstuff processing (Bonneau and Laarveld, 1999) and production of vaccines (Eloit, 1998), proteins, stem cells, tissues and monoclonal antibodies for use in therapeutics are not included here. The impact of reproductive technologies on animal breeding, not directly related with gene transfer, are reviewed elsewhere. The possible role for cloning adult animals in breeding is also discussed

Use of Genomic Information in Animal Improvement

The use of genomic information (sequences or DNA marker polymorphisms) for the genetic improvement and selection of animals requires the knowledge of the effect of physically mapped genes with effects on economically important traits or quantitative trait loci (QTL). This information is also required in order to effectively use transgenesis and MAS for genetic improvement. In MAS, the genomic information is combined with the classical performance records and genealogical information to increase selection accuracy, performing selection earlier in life and reducing costs.

The traits on which the application of marker-assisted selection can be more effective, are those that are expressed late in the life of the animal, have low heritability, are sex-limited, are expensive to measure or are controlled by a few genes. Examples are longevity, carcass traits in meat producing animals, and diseases or defects of simple inheritance.

Expected increments in selection response from MAS for a single complex trait, using known QTL genotypes plus linear model predictions (BLUP), compared to selection on BLUP alone, ranges from-0.7 to 64 percent. In practice, results will depend on many parameters which are likely to be very different for each trait combination and population. The statistical properties of genetic evaluations (predictions) of animals for quantitative traits obtained through mixed model methodology using phenotypic records and genealogical information as inputs are known as BLUP. Best-means minimum variance of prediction, Linear-because predictions are linear functions of observations, Unbiased-means that the expected value of predictors obtained with linear model have an expected value equal to the expected value of the mean of the breeding values, conditional to data, and Prediction-because involves prediction of random breeding values).

Most experiments on QTL detection in animals allow only the estimation of wide chromosomal regions (practical maximum resolution is of about 1 cM, but usual resolution is about 30 cM) that harbour a QTL in a 'statistical sense', estimated from the effects of some marker haplotypes on quantitative traits (de Koning et al. 2003). Thus, further confirmation is required in order to assure the use of the causative gene. Identification of the causative gene has proven to be difficult.

The process to identify the gene responsible for the effect is known either as 'fine mapping' studies (targeting mapping smaller genomic regions) or 'candidate gene' studies (targeting individual genes based on their probable function) (Lynch and Walsh, 1998). In practice, MAS is useful to select genes with effects well identified and precisely located in the genome such as those controlling monogenic recessive diseases such as the pig stress syndrome gene. However, for most recessive alleles with lethal or semi-lethal effects, natural selection will maintain their frequencies very low (Hartl and Clark, 1997) making MAS unnecessary.

Unless the additive and non-additive effects for most genes involved in the phenotypic expression of complex, economically important traits are determined, MAS should be regarded just as a tool to increase the rates of genetic gains and not a method to fully open the 'black box' of the genetic control of complex traits, that would render phenotypic selection 'obsolete'. Therefore, the perspectives on the optimum use of DNA marker information in the framework of a genetic program is still a matter of debate.

Quantitative trait loci experiments using crosses between breeds or lines with extreme genotypes for a trait, increases the power of detecting QTLs for that trait, compared to within-family designs. These across population's polymorphisms are not necessarily useful to perform MAS for within-population selection. The favourable allele could be fixed in parental populations and crosses may be commercially irrelevant. Wide genome scans for positioning a QTL using crosses or within-family experiments, are only the initial phase of the search for a true mayor gene involved in a complex trait (de Koning et al. 2003). Another source of complexity for detection and use of QTL for selection is genetic heterogeneity, where DNA mutations in several sites produce the same phenotype. Major single gene effects can be sometimes compensated in the organism using alternative metabolic pathways (McAfee, 2003).

Problems related to false positive detection of candidate genes are also common. Using crosses between two pig breeds, a polymorphism on the estrogen receptor locus (ESR) was associated to litter size in pigs with 1.5 piglet advantage for homozygous sows for the beneficial allele, and where followed by immediate recommendations for commercial use and patenting (Rotschild et al. 1996). Further research however did not confirm the effect.

Different phases of linkage between the markers and the QTL could explain the fact that the effect of the ESR locus varied widely between populations (Gibson et al. 2002). Thus, very probably, despite the ESR gene is probably a plausible 'candidate' from their inferred physiological functions (Rotschild et al. 1996), the gene involved seems to be another one, still unknown, or the effect initially observed was the product of several, interacting genes (epistasis).

Main problems related to the use of molecular genetics in the improvement of agricultural populations are:

1. Direct use of a discovered QTL effect for selection across families is not possible.
2. By the time the information about the inferred genotypes is known, frequently the animals involved in the study are not available as candidates for selection, because they will be too old.

3. Advantage from within-family selection for a QTL bracketed by markers over BLUP or phenotypic selection alone is frequently low and the methodology to exploit this information for selection is complex and relatively inefficient.
4. There are statistical estimation errors, causing both false positive and false negative effects, particularly when the effect of the QTL is small.
5. There is a lack of consistency of the effect of the same QTL between studies, caused by QTL by genetic background (epistasis) of QTL by environment interactions.
6. The net economic effect of the QTL may be lower than the effect on single traits, because unfavourable effects on other traits.
7. Selection using QTL is more complex than phenotypic selection alone. QTL information (whether the information on the QTL is direct or indirect), adds to the list of traits used as selection criteria. Issues such as reduction of selection intensities and relative emphasis given to each trait, make optimal selection more difficult, with a need for adequate relative weights for the QTL, and the polygenic portions of the genetic variation for each trait at each generation (year).
8. Short-term gains due to MAS may be at the expense of medium to long-term polygenic responses for important traits.

Even with an unambiguous knowledge for the allele effects of a mayor gene on a complex trait, expected advantages from optimum use of genotyping alleles for a QTL for a multi-generation selection horizon is not always high. The polymorphism for the αs1-casein in goats has a strong effect on protein content and total protein output. The difference between homozygous for the highest and lowest effects for milk protein is approximately three phenotypic standard deviations for milk protein content. Favourable alleles have frequencies lower to 0.5 in populations undergoing selection, making a very favourable case for potential gains in protein content and production from MAS using this polymorphism.

Simulation studies by Larzul et al. (1997), Fournet et al. (1997) and by Manfredi et al. (1998) indicated that when an efficient 'conventional' progeny testing selection program is underway for increased protein production, the advantages from MAS are low to moderate. Maximum possible increase on total genetic gain for protein yield was 26%. Dekkers and Hospital (2002) emphasized the overlap that exists between marker and phenotypic information for the improvement of a multi-trait goal over several generations, using MAS. A very optimistic prospect from use of MAS as well as other biotechnologies is very common in popular commercial and non-refereed publications, based on approaches based on exploiting single gene effects, without consideration to polygenic effects, economic values or time for fixation.

Research shows that the real situations are far more difficult for complex traits. These traits are controlled by several genes and environmental effects. Dekkers (2004) made a survey on the status of application of MAS in actual animal breeding programs for complex traits. He concluded that initial expectations for the use of MAS were high, but the current attitude is one of cautious optimism, with a need for careful examination of alternative selection strategies, business goals and integration of molecular and other technologies. Pollak (2005) made a detailed survey on the application of DNA technology for beef cattle improvement in USA. He concluded that current contribution of the new DNA technologies for beef cattle breeding is marginal, because they are encountering logistics and mechanical issues. For genomics technologies to impact fully on the beef industry, a higher level of sophistication of the genetic tests will be needed. Tests based on the genes themselves, rather than DNA markers associated with genes, will be required.

It is theoretically possible to predict accurately the breeding values of animals using many markers (Meuwissen et al. 2001). From this knowledge, it is possible to develop a model for *in vitro* genetic improvement of animals. This is known as velogenetics. The model involves *in vitro* selection of cells containing the desired genes the use of totipotent embryonic stem cells (ES). The procedure uses transfection of the desired genes, selection in vitro of the cells, and nuclear transfer of the desired genotypes into receptor oocytes. This approach is supposed to increase the rate of genetic improvement by obtaining many generations in a short time by avoiding rearing, reproduction and selection of 'real animals'.

Selection on the basis of genomic information only, such in this *in vitro* system, even with major genes with known effects well localized, may be dangerous, because in these artificial populations, unlike in real populations, natural selection would not be allowed to act at each generation on fitness traits under real, perhaps changing, environmental conditions. Changes on economically important traits will not be evaluated directly (Dekkers and Hospital, 2002). This may potentially reduce the responses on selected traits because of genotype x environment interactions (Montaldo, 2001). This is because selection is performed in artificial conditions that may deteriorate the fitness of the population and economic response.

Using MAS for improving health in animals by reducing disease prevalence (increasing disease resistance) or increasing resilience (the ability to withstand the disease without harmful effects), for infectious or parasitic diseases has been difficult. In most cases, excepting some rare examples such as Scrapie in sheep, complete resistance could not be obtained with the manipulation of a small number of genes. For most diseases, single-gene approaches are expected to have only a partial contribution. Gene interactions are common (Kuhnlein et al. 2003).

For many diseases, heritabilities are often low. That indicates the existence of many environmental factors affecting both the probability of infection and the response of the host. In spite of responses attained using conventional selection for some traits that are used as indicators of disease, the result is not well known. The existence of contradictory results regarding associations between production and disease resistance, the complexities of immune and resistance mechanisms and the interaction with other methods of control such as vaccination, sanitation, management and chemotherapy, makes the whole issue of selecting for disease resistance more difficult, in principle, that selecting for production traits. Moreover, we know that heritable resistance or resilience to more virulent form of pathogens would be increased by natural selection. As heritabilities for survival are generally low, we know that the genetic control of disease may be very complex, making difficult to change the outcome by manipulating single genes.

There is one published result on a successful MAS selection program to reduce the prevalence of dermatophilosis, a tropical infectious disease in Zebu cattle (Maillard et al. 2003). Maillard et al. (2003) argue to have obtained a sharp reduction in clinical prevalence of the disease from 0.76 to 0.02 in a period of five years by selecting against only two type II BoLA alleles associated with a high susceptibility of the disease. The authors explained the observed change resulting from selection performed in an unknown number of animals of each sex in 1996. However, a complete description of the changes in allele frequencies and genotypes from the moment of selection and their association with the evolution of prevalence by sex is not given. Considering the possibility of environmental changes and the presence of natural selection, in the absence of a control group, it is difficult to know if the observed change is the sole result of the mechanisms invoked by the authors through MAS.

We cannot at this moment forecast precisely the future of MAS in animal selection, but it is premature to conclude that methods based on phenotypic information will be replaced by methods based solely on genomic data. An integration of both types of data with the use of more sophisticated statistical models is needed. It is far from sure that total replacement of phenotypic information with gene-by-gene information, as selection criterion is possible or even desirable in the future.

Other very important applications of genetic markers in animal improvement include the optimization of mating strategies for non-additive genetic effects (estimation and managing of inbreeding and heterosis), parentage determination, genetic characterization of diverse animal breeds and populations using studies of between and within population (breeds) diversity (Oldenbroek, 1999) and marker-assisted introgression of particular alleles.

Cloning Adult Mammals

Cloning an animal is the production of a genetically identical individual, by transferring the nucleus of differentiated adult cells into an oocyte from which the nucleus has been removed. This is known as Nuclear Transfer and is how the Dolly sheep was produced. Since the publication of the original paper on cloning (Wilmut et al. 1997), there are several other reports on adult cloned animals involving mice, cattle, cats, goats, pigs, sheep and rabbits involving the same, and other cloning techniques.

Cloning Methods

In the case of Dolly, mammary gland cells in culture from a 6-year old donor ewe, where subjected to a reduction in the concentration of serum and thus obliged to enter in a quiescent state of the cell cycle (G0). Nuclear transfers to enucleated oocytes, was followed by electrical pulses for fusion of the donor cell nucleus and oocyte membranes and activate division (Wilmut et al. 1997).

Problems

Currently there is no doubts regarding the genetic similarity of the donor and the clone in the case of Dolly, however, besides low success rates (Edwards et al. 2003), several health problems related to the technique have been described (Samiec and Skrzyszowska, 2005). Normal development of an embryo is dependent on the methylation state of the DNA contributed by the sperm and egg and on the appropriate reconfiguration of the chromatin structure after fertilization. Somatic cells have very different chromatin structure to sperm and 'reprogramming' of the transferred nuclei must occur within a few hours of activation of reconstructed embryos. Incomplete or inappropriate reprogramming will lead to de-regulation of gene expression and failure of the embryo or foetus to develop normally or to non-fatal developmental abnormalities in those that survive. These facts indicate that there is a need for studies to determine further biological consequences of cloning. Cloning has important potential applications in gene transfer procedures.

Use of Cloning in Animal Breeding

Use of cloning in animal genetic improvement may increase the rates of selection progress in certain cases, particularly in situations where artificial insemination is not possible, such as in pastoral systems with ruminants. Currently, high costs of cloning are one of the main factors limiting their use as a technique in practical animal breeding. Clonal groups, however more uniform than full sibs, will have all differences caused by the environmental fraction of variation for measured traits, which is usually more than 50% of total variation. Selection among many cloned germlines allows the use of the non-additive genetic effects.

These effects are not exploited when traditional selection methods involving sexual reproduction are used in animal improvement (Visscher et al. 2000), but most of the observed genetic variation between animals is additive (Van Vleck, 1999). Advantages in terms of additional genetic progress however, seems to be only marginal from clone evaluation in selection nucleus herds (Ruane et al. 1997). Production based on clones of the best animals of the population, may allow for a one time large 'jump' in breeding value, so the commercial animals might be very close to those in the nucleus. However, further genetic improvement must be based in the continued use of the genetic variation by selection programs.

Transgenic Animals

Transgenesis is a procedure in which a gene or part of a gene from one individual is incorporated in the genome of another one. Transgenic animals have any of these genetic modifications with potential use in studying mechanisms of gene function, changing attributes of the animal in order to synthesize proteins of high value, create models for human disease or to improve productivity or disease resistance in animals. In the early 80´s, several research groups reported success in gene transfer and the development of transgenic mice. The definition of transgenic animal has been extended to include animals that result from the molecular manipulation of endogenous genomic DNA, including all techniques from DNA microinjection to embryonic stem (ES) cell transfer and 'knockout' mouse production (Cameron et al. 1994). Since the early 1980s, the production of transgenic mice by microinjection of DNA into the pronucleus of zygotes has been the most productive and widely used technique.

Using transgenic technology in the mouse, such as antisense RNA encoding transgenesis, it is now possible to add a new gene to the genome, increase the level of expression or change the tissue specificity of expression of a gene, and decrease the level of synthesis of a specific protein. Removal or alteration of an existing gene via homologous recombination required the use of ES cells and was limited to the mouse until the advent of nuclear transfer cloning procedures.

Transgenic Methods

Microinjection of DNA and now nuclear transfer, are two methods used to produce transgenic livestock successfully. The steps in the development of transgenic models are relatively straightforward. Once a specific fusion gene containing a promoter and the gene to be expressed has been cloned and characterized, sufficient quantities are isolated, purified and tested in cell culture if possible and readied for preliminary mammalian gene transfer experiments. In contrast with nuclear transfer studies, DNA microinjection experiments were first performed in the mouse (Izquierdo, 2001).

While the transgenic mouse model will not always identify likely phenotypic expression patterns in domestic animals, there have not been a single construct that would function in a pig when there was no evidence of transgene expression in mice. Preliminary experimentation in mice has been a crucial component of any gene transfer experiment in domestic animals (Kerr and Wellnitz, 2003). While nuclear transfer might be considered inefficient in its current form, major advances in experimental protocols, can be anticipated. The added possibility of gene targeting through nuclear transplantation opens up a host of applications, particularly with regard to the use of transgenic animals to produce human pharmaceuticals.

The only major technological advance since the initial production of transgenic farm animals has been the development of methods for the *in vitro* maturation of oocytes (IVM), *in vitro* fertilization (IVF) and subsequent culture of injected embryos prior to transfer to recipient females. Another highly efficient technique for transgenesis has been recently developed based on the use of lentiviral vectors to transform cow and pig oocytes. These vectors are more efficient than microinjection in terms of transformation and expression rates. One limitation is that the size of the transgene and the internal promoter has to be less than 8.5 kb in size.

Transgenesis in the Improvement of Production traits

The technology of transgenesis is potentially useful to modify characters of economic importance in a rapid and precise way. Contrary to the 'classical' selection programs, it is necessary a knowledge of the genes that control these characters and their regulation.

Following is a brief discussion of experiences with transgenesis to alter economically important traits in livestock.

Growth and Meat Traits

In most of the earlier work in domestic species (pig, sheep, rabbit) growth hormone was enhanced by the metallotionein promoter to control its expression. Subsequent efforts to genetically alter growth rates and patterns have included production of transgenic swine and cattle expressing a foreign c-ski oncogene, which targets skeletal muscle, and studies of growth in lines of mice and sheep that separately express transgenes encoding growth hormone-releasing factor (GRF) or insulin-like growth factor I (IGF-I). Transgenic pigs and sheep with high levels of serum growth hormone were obtained, but an increment of its rate of growth was not observed, and only in some lines average daily gain increased with the supplement of the diet with high levels of protein. The highest effects were observed in the reduction of body fat. A large number of different serious pathologies and a severe reduction in reproductive capacity were described in these animals (Murray et al. 1999). In a report about two studies with pigs (Neimann, 1998), there is evidence for the use of transgenesis allowing to important reductions in body fat and increased diameter of muscle fiber by increased IGF-I levels and growth hormone without serious pathological side effects. Australian regulations avoided the commercial release of these animals.

Frequently the used promoters have not allowed an efficient control of the expression of the transgene. It was assessed that it is necessary to develop more complex constructions that activate or repress the expression of the transgene more precisely. Adams et al. (2002) found inconsistent results regarding the effect of a growth hormone construct in sheep on growth and meat quality.

Recently, a spectacular transformation was obtained by insertion of a plant gene in pigs. Saeki et al. (2004) generated transgenic pigs that carried the fatty acid desaturation 2 gene for a 12 fatty acid desaturase from spinach. Levels of linoleic acid (18:2n-6) in adipocytes that had differentiated in vitro from cells derived from the transgenic pigs were 10 times higher than those from wild-type pigs. In addition, the white adipose tissue of transgenic pigs contained 20% more linoleic acid (18:2n-6) than that of wild-type pigs. These results demonstrate the functional expression of a plant gene for a fatty acid desaturase in mammals, opening up the possibility of modifying the fatty acid composition of products from domestic animals by transgenic technology.

Wool Production

The objectives are to improve production of sheep wool and to modify the properties of the fiber. Because cystein seems to be the limiting amino acid for wool synthesis, the first approach was to increase its production through transfer of cystein biosynthesis from bacterial genes to sheep genome (Murray et al. 1999). This approach did not achieve the efficient expression of these enzymes in the rumen of transgenic sheep.

Milk Composition

Milk proteins are coded by unique copy genes that can be altered to modify milk composition and properties. Among the different applications of milk modification in transgenic animals, the following can be highlighted:

1. To modify bovine milk to make it more appropriate to the consumption of infants. Human milk lacks β-lactoglobulin, has a higher relationship of serum proteins to caseins, and has a higher content in lactoferrin and lysozyme when compared to bovine milk. Lactoferrin is responsible for the iron transport and inhibits the bacterial growth. To introduce the human lactoferrin into the bovine milk, transgenic cows have been obtained. The elimination of the β-lactoglobulin in the cow milk would be another interesting objective because is one of the major allergens of cow's milk.
2. To reduce the content of lactose in the milk to allow their consumption to people with intolerance to lactose. It is considered that 70% of the world population is lacking theintestinal lactase, the enzyme required to digest the lactose. The reduction in lactose may be obtained by expressing β-galactosidase in the milk or

diminishing the content of α-lactalbumin. Transgenic mouse with inactivated α-lactalbumin gene produce milk without lactose. However, a serious practical drawback of this method is that this milk is very viscous and it is not secreted to the exterior of the mammary gland, due to the importance of the lactose in the osmoregulation of the milk (Stinnakre et al. 1994).

3. To alter the content of caseins of the milk to increase their nutritive value, cheese yield and processing properties. Research has intended to increase the number of copies of the gene of the κ-casein, to reduce the size of the micelles and modificating the κ-casein to make it more susceptible to the digestion with chymosin. This has only been done using the mouse as a model (Gutiérrez-Adan et al. 1996). Brophy et al.(2003) engineered female bovine foetal fibroblasts to express additional copies of transgenes encoding two types of casein: bovine β-casein and κ-casein. The modified cell lines of fibroblasts were used to create eleven cloned calves. Milk from the cloned animals was enriched for β-and κ-casein, resulting in a 30% increase in the total milk casein or a 13% increase in total milk protein, demonstrating the potential of this technology to make modified milk.
4. To express antibacterial substances in the milk, such as proteases to increase mastitis resistance. The objective is to alter the concentrations of antibacterial proteins such as lyzozyme or transferrin in the milk.

Future Perspectives of Transgenesis

The techniques for obtaining transgenic animals in species of agricultural interest are still inefficient. Some approaches that may overcome this problem are based on cloning strategies. Using these techniques it is feasible to reduce to less than 50% the number of embryo receptor females, which is one of the most important economic limiting factor in domestic species. It would also facilitate the further proliferation of transgenic animals. Recent results relate these techniques with still low success rates (Edwards et al. 2003), high rates of perinatal mortality and variable transgenic expression that requires to be evaluated before generalizing their application.

Considerable effort and time is required to propagate the transgenic animal genetics into commercial dairy herds. Rapid dissemination of the genetics of the parental animals by nuclear transfer could result in the generation of mini-herds in two to three years. However, the existing inefficiencies in nuclear transfer make this a difficult undertaking. It is noteworthy that the genetic merit of the 'cloned' animals will be fixed, while continuous genetic improvements will be introduced in commercial herds by using artificial insemination breeding programs (Karatzas, 2003).

In an alternative scenario of herd expansion, semen homozygous for the transgene may be available in four to five years. Extensive breeding programs will be critical in studying the interaction and co-adaptation of the transgene(s), with the background polygenes controlling milk production and composition. Controlling inbreeding and confirming the absence of deleterious traits so that the immediate genetic variability introduced by transgenesis is transformed into the greatest possible genetic progress is equally critical (Karatzas, 2003). Another alternative strategy for transgenesis is based on the use of sperms as vectors in the integration of the transgenes. Initially described in mice (Lavitrano et al. 1989). Results showed that this procedure might be efficient in sheep (Niemann, 1998). In addition, a successful expression of a gene related to genetic modification of pigs for a gene related to xenotranplantation was obtained using this technique.

Eighty percent of the pigs were transformed and 54% expressed the transgene consistently (Lavitrano et al. 2002). A very efficient modification of this technique that uses the co-injection of sperms and DNA, has been described in the mouse and given a high rate of transgenesis (20%), therefore, their application to domestic species seems promising. Intracytoplasmic sperm injection (ICSI) has been used recently for the stable incorporation and phenotypic expression of large yeast artificial chromosome (YAC) constructs of submegabase and megabase magnitude. This technique allowed for more than 35% of transgenesis (Moreira et al. 2004).

Another option for transgenesis is the use of insertional mutagenesis using natural transposons. A transposon system called "Sleeping Beauty", and active in a wide range of vertebrate cells, was used to transform mouse embryos with mRNA expressing the SB10 transposes enzyme (Dupuy et al. 2002). Kuroiwa et al. (2004) targeted sequentially a system for primary fibroblasts cells that were used to knock out both alleles of a silent gene, the bovine gene encoding immunoglobulin-μ (IGHM), and the active gene encoding the bovine prion protein (*PRNP*) and produced both heterozygous and homozygous knockout calves. The procedure integrates homologous recombination to replace genes in cell culture, and rejuvenation of cell lines by production of cloned fetuses.

A method for selective elimination of selection marker genes was also developed. This method allow for the production of double homozygous transgenic embryos in 21.5 months. In contrast, for cattle, the production of double homozygous from heterozygous founders would require approximately 5 years and generation for double homozygous from heterozygous founders is impractical. This method can be used to breed many types of cattle with improved disease resistance and values for increased productivity. A recent alternative consists on the transformation of somatic tissues of developed animals, using techniques similar to those used in gene therapy (Kinghorn, 2003).

Discussion

Detecting genes related to disease and their expression in humans from studies on the genome, could lead to the development of therapies and the development of drugs for specific individuals, and enhanced early diagnosis of individuals with high-risk genotypes, allowing for preventive or remedial actions, even gene therapy. In animals, this knowledge could lead, in addition, to select against defective genes.

In livestock, knowledge of effects of specific genes and gene combinations on important traits could lead to their enhanced control to create new, more useful populations. The use of specific gene information is not a panacea, but could help to increase rates of genetic improvement, and open opportunities for using additive and non-additive genetic effects of domestic species, provided wise improvement goals are used and this new technology is optimally used together with the so called 'traditional' or 'conventional' methods based on phenotypic and genealogical information.

These methods will help to increase our knowledge about the genetic architecture of complex quantitative traits in domestic animal populations and to estimate the distribution of the genetic variation across and within breeds and population. It will also aid in ascertaining the genetic merit of local, less known populations (Hill, 2000). Studies for using genetic diversity in structured populations using DNA markers (Hartl and Clark, 1997) are very useful in order to set priorities for conservation of distant or unique populations as reservoirs of potentially unique genes, because their contribution to biodiversity would be greater (Oldenbroek, 1999). Currently, however, the main practical application of DNA markers is for parenting determination and to trace products such as meat. Despite its relatively low success rates and associated high costs, transgenic technology have a number of important potential applications in animal improvement such as increasing productivity, product quality and creating novel products.

A major limitation to use transgenesis in the improvement of productive characters is the limited knowledge available on the identity and regulation of the genes that control these characters. The advance in the elaboration of genetic maps and fine positional cloning studies in the main species of interest will allow having a larger number of candidate genes susceptible of being manipulated. However, the road from genotype to phenotype is proving to be much more complex than previously thought for disease and production traits affected by many genes (True et al. 2004).

One promising applications of transgenesis is the synthesis of biomedical products of high commercial interest. Transgenic bioreactors and the use of exogenous or artificial genes interfering with particular cell mechanisms or with pathogens but not, or only marginally, with the physiology of the animals are potential applications. A greater knowledge on the mechanisms that determine the integration of the transgenes and genic regulation will allow a more precise control of the expression of the transgenes and it will probably facilitate a larger number of applications in the domestic species, including modifications beyond normal limits, such as to increase the number of copies of the gene and their expression.

These transformations could be regarded as a form of mutation (Hill, 2000). The expressions of complex traits are the result of several mechanisms involving both regulatory and structural portions of the genome. Advances in molecular genetics, genomics, proteomics and transcriptomics might perhaps help to shorten the gap between the more 'holistic' approaches of quantitative genetics with the more 'reductionistic' approach of molecular genetics. The release of genome sequence information in cattle (Sonstegard and Van Tassell, 2004) and pig (Wernersson et al. 2005), may allow for a more efficient use of MAS and also to address some consumers concerns regarding product quality and safety.

Use of genetic engineering for animal and plant improvement is in its infancy, therefore many questions regarding efficiency, safety and societal benefits in particular situations remain. Problems arising transgenic plants, including their lower-than expected productivity, are reviewed thoroughly by McAfee (2003). Simplistic and overoptimistic views of biotechnology should be replaced by serious and scientifically based assessments of these new technologies by potential users on a case-by-case basis. We need to emphasize that in most cases, the use of MAS is not a revolution but just an evolution with regard to the traditional methods, because we are looking to improve more efficiently traits that already are actually or potentially improved in an efficient way using, for instance, mixed model (BLUP) based technologies for selection.

Efficiency issues are very important. In order to increasing the efficiency of MAS, we need previously to:

1. Define with greater precision the selection goal and selection criteria (Monin, 2003).
2. Optimize the use of BLUP and other 'classical' breeding methodology.

The use of transgenic animals models for the study of gene regulation and expression has become commonplace in the biological sciences. Contrary to the early prospects related to commercial exploitation in agriculture, there are some challenges regarding their use that still lay ahead.

The risks at hand can be defined not only by scientific evidence but also in relation to public concern (whether perceived or real) that exists in some people (Larrère, 2003). Therefore, the central questions will revolve around the proper safeguards to employ and the development of a coherent and unified regulation of the technology.

Cloning is another technique that raises concerns both from the ethical and practical point of view. Whether it is acceptable to clone humans is a very difficult issue. In animals, besides the very low success rates, some abnormalities should suggest that more information is required on the consequences of such practices in humans but also in animals, before its routine use. Advantages for animal breeding programs derived from cloning with no use of transgenesis are like to be small (Van Vleck, 1999).

These two examples illustrate that in spite most of the problems are technical in nature, implications of the use of this knowledge will be important for the society as a whole (Olsson and Sandoe, 2004).

A reasonable degree of regulation, open information on the issues of genetic engineering technologies from the academic world and an involvement of the whole society in the developments of the laws concerning these issues, seems to be the best way to circumvent an exaggerated or negative reactions to some of these knowledge, and to avoid or reduce unethical or abusive use of these techniques (Fukuyama and Stock, 2002).

A specific set of conclusions regarding safety of food from genetically modified animals is available from a FAO/WHO expert consultation panel (FAO/WHO, 2003).

Concluding Remarks

Most of the important potential technical advances offered by genetic engineering technology in animal breeding are still ahead. Their use has both advantages and problems. Advantages are related to a more complete control over the animal genome. Problems are related to technical complexity, high costs, in some cases, public acceptance and ethical dilemmas. It is not likely that this technology, will replace 'conventional' methods for genetic improvement. Instead, they probably will begin to be gradually incorporated into current genetic improvement programs that use efficiently classical improvement methods to achieve particular objectives.

Bioland Standards

"No act contrary to nature remains without consequences. No natural principle can be breached without its being punished, no natural order of things be dispensed with without danger to ourselves. The integration of humans in the order of creation is a vital prerequisite for their lives." —Dr. H. P. Rusch

Dr. Hans Müller and Dr. Hans Peter Rusch, in their work on the care of the soil and the maintenance of its long-term fertility, established the organic biological method of farming. This is based on the exact observation of biological connections of the effects between soil – plants – animals and humans with the aim of achieving optimum care of biological regulation systems in the agricultural field. Agricultural products are generated within as closed a business operating cycle as possible in the sense of a true original production. The mutual tasks of organic biological cultivation consist of:

- caring for the natural basics of the life of the soil, water and air
- producing foodstuffs of a high health value
- carrying out active nature protection and the preservation of species
- avoiding to damage the environment
- keeping animals according to the needs of its species
- making a contribution towards solving the world-wide energy and raw materials problems
- creating the basis for the maintenance and development of independent farming structures.

For decades farmers have been working according to the knowledge gained by Dr. Müller and Dr. Rusch and have mutually developed this further in their practical work. It has thus been possible for them in their fields of work to counteract the negative effects of the agricultural and social politics, to operate an environmentally friendly form of agriculture and, in co-operation with processors and consumers, to put a stop to the destruction of the existence of farmers. These farmers, gardeners, wine-growers and beekeepers in the Federal Republic of Germany combined to establish the BIOLAND e.V. Verband für organisch-biologischen Landbaumethoden (further in the text called BIOLAND) and have compiled the following standards.

The standards explain in detail the application of the organic biological methods of farming, the conversion to this method of operation and enable control of the cultivation defined according to the standards to be executed. It remains the mutual task of the people connected with BIOLAND to continue to work towards the aim of maintaining our natural basics of life and to improve the standards to keep them in line with the latest knowledge available.

EU-Regulation "Organic Agriculture"

During the drafting process of these standards the "EU Regulation on Organic Production of agricultural Products and Indications Referring thereto on Agricultural Products and Foodstuffs (EEC) Nr. 2092/91" and its amendments have been observed. BIOLAND member farms and contract partners are always obligated to adhere to the provisions of this EU-regulation in its currently amended form.

Genetic Engineering

Exclusion of Genetic Engineering

Genetically modified organisms (GMOs) and products derived therefrom are not compatible with the organic production method. Products, produced according to BIOLAND-standards, have to be produced without the use of genetically modified organisms (GMOs) and/or GMO derivatives.

Definition of Terms

'Genetically modified organism' (GMO) shall mean any organism as defined in Article 2 of Council Directive 90/220/EEC of 23 April 1990 on the deliberate release into the environment of genetically modified organisms 'GMO derivative' shall mean any substance which is either produced from or produced by GMOs, but does not contain them. 'Use of GMOs and GMO derivatives' shall mean use thereof as food stuffs, food ingredients (including additives and flavourings), processing aids (including extraction solvents), feeding stuffs, compound feeding stuffs, feed materials, feed additives, processing aids for feeding stuffs, certain products used in animal nutrition (under Directive 82/471/EEC), plant protection products, veterinary medicinal products, fertilisers, soil conditioners, seeds, vegetative reproductive material and livestock.

Crop production

Soil Fertility

The care of the soil and, correspondingly, the maintenance and the improvement of soil fertility constitutes a special point of emphasis in organic biological farming. A healthy, invigorated soil is the best prerequisite for healthy plants, healthy animals and healthy people. All measures of plant growing should form the basis for the improvement and care of a diverse and active soil life. Only the vitality of the soil itself will ensure long-lasting fertility.

Location

Ecological Design

In order to promote the health and the resistance of plants, the location must be designed in accordance with ecological points of view. For example, by planting and maintaining hedges, creating nesting possibilities and ensuring provision of shelter for insects, beneficial animals are to be encouraged and the self-regulation within the ecological system improved.

Selection of Location

In the choice of the location, the load created by pollutants from the environment and from previous usage of the soil are to be taken into consideration. If there is the danger of such a load being present, food stuffs and soil must be examined for residue.

Areas which have been affected by loads can only be used for organic biological farming when the loads involved have been reduced to a level which is justifiable for health by the adoption of suitable measures (e.g. protective planting). The BIOLAND Association can prohibit the use of the Association's name/trade mark BIOLAND on products which have been produced from land, partial land or border land contaminated by such loads.

Crop Rotation

Crop rotation is to be planned in such a variable and balanced manner that this fulfils the following functions:

- the maintenance of soil fertility
- the production of healthy plants
- the suppression of weed in fields
- the nutrition of animals using the business's own fodder
- the achievement of economically feasible yields without the use of chemical fertilisers and chemical products for plant protection.

In order to fulfil these functions, crop rotation must contain leguminous plants as main or intermediate crops or as mixed cultures.

Soil Preparation

The objective of soil preparation is the creation of optimum growth conditions for the crops. The compatibility with the soil life is to be taken into consideration in all measures adopted in soil preparation. Soil preparation must be carried out in such a manner that the natural soil structure is not excessively disturbed and that a loss of nutritional content and unnecessary expenditure of energy are avoided.

Fertilisation and Humus Management

Basic Principles

The objective of fertilisation is to achieve harmonic nutrition of the plants by means of a soil full of life. Organic material from the business itself forms the basis of fertilisation. It is mainly added to the soil by means of spread composting. Manure from the business itself must be prepared and spread in such way that the life in the soil is supported and the humus content is maintained or increased.

Permissible External Fertilisers

In order to complement the fertiliser produced in the business itself and to compensate any losses in nutrition caused by the operational cycle, fertilisers from external farms and organic and mineral fertilisers may be used in as far as these are listed in 10.1. Farm fertilisers from conventional sources must be subjected to careful composting.

They may only be used when they are considered harmless in regard to their pollution content. If necessary a quality examination can be requested. Trace elements may only be used when the deficiency determined cannot be removed by any other means.

Non-permissible Fertilisers

The use of farmyard slurry and urine and poultry manure from conventional animal farming is forbidden. In addition, the use of chemical synthetic nitrogenous fertilisers, easily soluble phosphates and other fertilisers not listed in 10.1 is prohibited.

Quantity Limitation

The total volume of organic fertiliser, based on the nitrogen content, may not exceed the amount which corresponds to an animal livestock count of 1.4 manure units (= DE) per ha. A maximum of 0.5 DE of this may be organic fertiliser from external sources.(DE = maximum animal stock density according 1.4 DE). The conditions specifie apply to gardening and perennial crops. In measuring the fertilising, the reserves available in the soil must be taken into consideration.

Production of Quality and Environmental Compatibility

Fertilising is to be designed in conformity with the location and the crops involved in such a way that the quality of the products (physiological nutritional value, taste, imperishability) may not be detrimentally affected in particular by the amount of nitrogenous fertiliser. In regard to the type, the amount and the time of applying fertiliser, care must be taken to avoid placing loads on the soil and the water (e.g. through heavy metals and nitrates).

Sewage Sludge and Compost

The use of sewage sludge and refuse compost is prohibited. Organic compost from separated collection and peat substitutes (e.g. bark products) may only be used after prior analysis of their pollutant content and following agreement with the BIOLAND Association.

Seeds, Seedlings and Plant Materials

For growing, those species and varieties of plants should be used which are best suited for the conditions prevailing at the location, they should not easily be subject to disease and be of a high physiological nutritional quality. In farming, varieties typical for the area should be used in preference to hybrid varieties.

Organically Produced Seeds and Plant Materials

When certified seeds and plant materials of suitable varieties are available from organic propagation, then these must be used. Any other sources require the express exceptional approval by the BIOLAND Association. From August 1st, 2006 on it is the objective to use seeds and plant materials from organic propagation exclusively.

Treatment of Seeds

Seeds and plant materials may not be treated after the harvest with chemical synthetic pesticides (e.g. disinfectants). Care is to be taken when using conditioned seeds (pelleted seeds, seed plates, etc.) to ensure that the materials used are harmless in the sense of these standards.

Seedlings

The seedlings used in the business must be grown by the business itself or be purchased from other farms of the BIOLAND Association, if here not available in accordance with the requirements of BIOLAND from other organically managed farms Substrates for cultivation may only contain a maximum of 80 vol. % of peat. Peat substitutes must low in pollution and ecologically compatible.

Young Plants for Perennial Crops

Young plants used in the business must be purchased from nurseries or producers of propagating materials of the BIOLAND Association, if here not available in accordance with the requirements of BIOLAND from other organically managed farms, if the required varieties and suitable qualities are available. Other sources require express approval by the BIOLAND association.

Plant Protection

The objective of organic-biological farming is to produce plants under such conditions that their infestation with parasites and disease achieves a point where this is of no or only minor economic significance. Appropriate measures for the achievement of this are balanced crop rotation, selection of suitable varieties, soil preparation in accordance with the location and the time of year, fertilising in appropriate amounts and qualities, fertilising by growing, etc. In addition, the spread of beneficial animals should be promoted by suitable means and measures such as hedges, nesting places, wet biotopes, etc. If equipment is used that is also used to disperse non-permissible plant protection agents according to the Bioland standards, prior to the use it has to be drained and cleaned thoroughly.

Permissible Measures

Special preventive measures should only be carried out using the agents which are listed in They are only to be used when all other measures for activating the defensive powers of the soil and the plants themselves and the design of the location have been exhausted. The legal regulations regarding their use are to be observed in using plant treatment agents.

Prohibitions

The use of synthetic pesticides and growth regulators is forbidden.

Weed Regulation

The regulation of weeds is effected by preventive measures (e.g. crop rotation, soil preparation, variety selection), mechanical measures (e.g. harrowing, raking, hoeing) and thermal measures (e.g. burning off).

Prohibition of Herbicides

The use of herbicides is forbidden.

Air, Soil and Water Protection

Water resources are not to be used excessively. It is not permissible to burn used plastic (e.g. foils and fleeces) in the fields.

Wild Collection

The collection of edible plants or parts thereof, growing naturally in natural areas or forests, where the only humen interference consists of the hearvest (collecting) of the products, is considered as wild collection, if the following conditions are observed:

- The collecting area must be clearly defined. It must be identified by the way of land register maps (if necessary drawing of plans).
- The collection in areas out of the area under the care of the BIOLAND Association is only allowed with prior approval.
- The collecting area shall not be under the direct influence of any sources of pollution.
- The areas have received no treatments with products other than those allowed by this standards (Annex 10.1 and 10.2) for a periode of three years prior to the collection. This must be documented by appropriate means.
- The collection shall not affect the stability of the natural habitat or the maintenance of the species in the collection area.

Those products may be labelled with the trade mark/association name of BIOLAND with the addition "... from wild collection" (processed products in the list of ingrediences).

Animal Husbandry

Keeping animals is a sensible link in the operational cycle. Keeping animals in accordance with the needs of the species and their considerate care by humans form the prerequisite for the health, the performance and the well-being of the animals. With the help of the animals, the feed produced in the operation is used in the production of foodstuffs of a high quality for human consumption. Animal keeping is to be designed in such a way that it can be assured that the production, storage and spreading of the manures produced in the operation of business through keeping animals is with as little loss as possible. This serves to maintain and improve the fertility of the soil in the business.

Requirements in the Keeping of Animals

Keeping animals in accordance with the needs of the species must be the objective of every business. This means that the behaviour peculiar to the species in question, such as behaviour in movement, rest, ingestion of feed, social contact, comfort and reproduction is made possible as far as feasible. To promote robustness and vitality, the animals should often be allowed to face the weather and climatic conditions of the location.

Keeping animals in a manner peculiar to the species involves providing sufficient space for movement and rest throughout the year, natural light, shade, protection against wind, fresh air and fresh water. It is mandatory for all animals to have access to open air and/or grazing (for existing livestock buildings transitional periods until 2010 are possible for poultry, pigs and young cattle with permission of the BIOLAND Association, compare 9.4). Injuries and illnesses resulting from their being kept must be avoided. Herd animals may not be kept individually. Keeping animals individually is permissible only in case of male animals for breeding purposes, in case of sickness, versus the end of pregnancy and in small stocks.

Space Requirements

Requirements for in-and outdoor areas of the livestock housing system are listed for each type of animal in attachment no. 10.6 (for existing livestock buildings transitional periods until 2010 are possible with permission of the BIOLAND Association, compare 9.4). For the keeping of fallow-deer and red deer the regulations. Housing systems for mammals with no clear separation between in-and outdoor area have to fulfil space requirements in sum. In case of housing systems for ruminants and horses with free range barn and access to pasture in summer, the space requirements for outdoor area according to attachment 10.6 can be skipped. In this case permanently accessible, durable and non-roofed barn area can be taken into calculation of the total barn area.

In regions with suitable climatic conditions, allowing animals to be kept outdoors all year round, housing is not stipulated.

Movement and Rest Area

Barns with fully perforated floor area (fully slatted floors, flat decks, cages) are not permissible. The width of slots and holes in case of perforated floors have to be adapted to animal size. Slatted floors must be in excellent technical condition. Surface slats are to be preferred. The majority of accessible movement and rest area for each mammal category must be a solid floor area (no slat floors). The tread surface must be non-slip and of tread proof nature.

A soft, dry and clean place in which to rest is to be ensured at all times for ruminants, pigs, horses and rabbits by means of strewing (as a rule, straw). Straw for bedding purposes should, as far as available, be from the farm itself or from other organic farms. Conventional straw for strewing should be grown on lands with a minor degree of farming intensity.

Assessment of Housing Systems

As an orientation aid in evaluating the adequacy of the compliance in the keeping of animals, the index of adequacy in animal keeping (TGI = Tiergerechtigkeitsindex, prepared by the society for ecological animal keeping (GÖD = Gesellschaft für ökologische Tierhaltung)) can be applied. New buildings and alterations to existing buildings for keeping animals should be in accordance with the latest status of knowledge in regard to keeping animals in compliance with the needs of their species. The planning of such buildings should therefore be co-ordinated with the BIOLAND Association.

Access and Care of Open Air Run Areas

Access to open air run and pasture has to be afforded always when physiological, climatic and soil conditions allow for this. The stock density of animals on outdoor areas may not result in the soil being trampled down – with the exception of feeding and drinking areas. Over-grazing has to be avoided.

Construction and Maintenance of Livestock Buildings

In construction and maintenance of livestock buildings ecological issues have to be considered. Substances hazardous to health and environment in building materials and its treatments have to be avoided, if possible.

Native building materials have to be preferred.

The use of non-regenerative energy resources in the construction and maintenance of stables has to be reduced, if possible.

Keeping Cattle

Dairy Cattle and Suckler Cow Keeping: Cows are afforded access to meadows or are allowed out into the open air at least during the six summer months.

Cows should be afforded the possibility to calve in separated calving booths or on the pasture.

Non-Penned Cowsheds

Efforts should be made to have non-penned cowsheds which allow the cattle freedom of movement.

Dead ends and bottle necks in the non-penned cowsheds are to be avoided.

If outdoor grazing in summer is not possible, then access to open air has to be afforded all year round (for existing livestock buildings a transitional period until 2010 is possible with the permission of the BIOLAND Association, compare 9.4).

In winter the possibility of regular movement in the open air should also be afforded. There must be a place in the non-penned sheds for each animal to sleep and eat. A slight reduction in number of eating places is possible with the permission of the BIOLAND Association in case of permanent availability of fodder (storage feeding system).

Boxes in which the animals can rest must enable the animal to lie down and rise up in a manner in compliance with the species. Keeping animals tethered in combination with summer pasture and/or regular access to open air is possible with the permission of the BIOLAND Association during a transitional period until

2010 in existing livestock housing (compare 9.4). Tethering system exceeding 2010 is possible for small animal stocks if cows are afforded access to open air, at least twice a week. Tethering of single animals for security or animal protection reason is possible with the permission of the BIOLAND Association as long as it is limited in time. If the animals are kept tethered, the width, the length and the technology used in tethering and thedesign of the edges of the trough must allow the animal to stand up, lie down or eat in a manner suitable for the species and must allow the animal sufficient body care.

The cows must be able to fully stand and rest on the level, secure surface, which has to be strewn sufficiently. Rigid neck frames and tightly drawn chains or nylon belts are not permissible. Electric cattle trainers are forbidden.

Cattle for Breeding and Beef Cattle

All cattle for breeding purposes and for beef should have the possibility of free movement throughout the whole year. Efforts should be made for grazing throughout the vegetation period. If grazing in pastures is not possible, cattle for breeding and beef can be kept throughout the year in non-penned cowsheds (for existing buildings a transitional period until 2010 is possible with the permission of the BIOLAND Association, compare 9.4). Keeping beef cattle without access to open air is permissible only during the end of the fattening period to a maximum of 1/5 of the life time, but in any case no longer than 3 months. If grazing possibilities are afforded during the vegetation period, tethering is permissible for breeding and beef cattle of an age of over 1 year.

Calves

The calves should be able to stay with the mother for at least 1 day following the birth. From their second week of life, when the number kept is correspondingly large, the calves must be kept in groups. In case of their being kept in cowsheds (huts/igloos) with the respective generous possibility to move about and to have social and eye contact, they can be kept in groups from the 6th week of life on. Tethering of calves and young cattle under the age of 1 year is not permissible.

Keeping Pigs

Pigs must be allowed access to open air runs (for existing livestock buildings transitional periods until 2010 are possible with the permission of the BIOLAND Association). With the exception of the late pregnancy period and the suckling period of sows, pigs have to be kept in groups. Tethering of sows is not permissible. Fixing should only be undertaken with problematic animals during and after farrowing. A wallowing area should be available. Housing pigs without access to open air is only permissible during the final fattening period for a maximum of 1/5 of life time, but in any case no longer than 3 months. During the 6 summer months period breeding pigs, wherever possible, are to be afforded access to a pasture. The pasture should have shady areas and an area for wallowing.

Sheep and Goats

The stables must be designed in a non-penned manner. During vegetation period efforts must be made to keep them on pastures. If no pastures are afforded, sheep and goats have to be kept in non-penned stables with access to open air. During final fattening period the keeping of fattening lambs and goats is only permissible to a maximum of 1/5 of the life time, but in any case no longer than 3 months.

Barn

Keeping laying hens in barns is in the form of floor or volary systems with access to open air run. The single barns with a maximum of 3000 laying hens have to be separated in such a way, that infections and/or a contamination with parasites will be reduced, and to ensure a sustainable management of the greened area for movement (for existing barns transitional periods until 2010 are possible with the permission of the BIOLAND Association, compare 9.4). Per each m^2 of movement area in the barn accessible to the animals, 6 animals can be kept. Movement area that accounts for the calculation of the animal stock density has to fulfil the following requirements:

- minimum width at least 30 cm
- maximum slope 40 %
- in case of gritted floor minimum wire strength to be kept at 2 mm
- free height between floor levels or perch rods at least 45 cm
- durable floor area has to be covered with suitable strewing material in sufficient thickness
- laying nests, its landing grids and higher perch rods are no moving areas and can thus not be accounted for in animal density calculations.

The exterior climate area can be accounted into accessible barn area, if:

- it is accessible through all barn openings for all animals during their total activity time (light phase, natural and artificial light)
- if it is roofed and equipped with automatic opening system, lighting, fencing and wind protection possibility (only in very cool temperatures and strong winds the number of barn openings can be reduced)
- if the whole exterior area is strewn with sand or similar for all animals
- if it has a height of at least 2 m
- if it is located on the same level as the barn; a level difference between barn and exterior area is limited to a maximum of 50 cm (in case of higher level differences, a sufficient circulation of animals can be reached by the construction of balconies and climbing and descending supports).

In relation to the barn floor area the maximum animal density in volary systems is 12 animals/m^2. In barns with an integrated exterior climate area, a maximum of 8 laying hens/m^2 accessible area in exterior barn areas can be kept at night, in volary systems a maximum 15 animals/m^2 floor area. The barn has to be designed in such a way that animals will have the least possible contact with excrement. The different floor levels accessible to laying hens have to be arranged in such a way that excrement do not fall on the level beneath it. At least 1/3 of the moving area in the barn for all animals must be available as strewn area for the purpose of scratching. In barns with integrated exterior climate area this third refers to the interior area of the barn. The strewing material has to be at least 5 cm deep and must be kept dry, loose and clean. The barn has to be lighted sufficiently with day light. The window area has to be equivalent to at least 5 % of the barn floor area. Natural day light may be extended to a maximum of 16 hours by artificial light. The offered space for feeding, feeding dishes and the strewn area for the application of feeding seeds have to be designed in such a way, that all animals can forage together.

The animals should be able to take water from an open water surface. The available drinking water has to be always fresh and clean. At least 18 cm of perch rod have to be provided. In barns with excrement pits at least 1/3 of the perch rods have to be elevated at least 45 cm. The profile of the perch rods must have at least 30 x 30 mm, the upper edges of the rods have to be rounded. For the total perch rod length only such perch rods are accounted for, that have at least 30 cm horizontal axis distance from each other and at least 20 cm distance from the walls. For the laying of the eggs the animals must have available sufficient strewn laying nests or rolling nests with smooth rubber nops or similar material. For 80 laying hens 1 m^2 family nest has to be available, a single nest is sufficient for a maximum of 5 hens.

The animals must have permanent access to a dust bath, if possible in a winter garden. The barn openings to the exterior climate area and the access to the outdoors are to be measured in such a way that the animals can circulate without problems and unrestricted. The barn openings have to add up to a combined length of 4 m per 100 m^2 of the floor area of the building available to the hens (for existing buildings a transitional period until 2010 is possible with the permission of the BIOLAND Association, compare 9.4). The minimum measures of the openings are 50 cm width and 45 cm free height. Between charging the barn has to be cleaned and disinfected. Only the substances listed in attachment 10.7 are permissible.

Exterior Climate Area (Winter Garden)

For an animal stock density of more than 4 hens per m^2 in the barn at least 1 m^2 of durable exterior climate area per 12 hens is mandatory. Excepted from this are stocks of less than 200 hens and mobile barns. The exterior climate area can be accounted for in the calculation of the animal density provided they fulfil the conditions.

Green Open Air Run

A green open air run is mandatory (for existing buildings transitional periods until 2010 are possible with the permission of the BIOLAND Association, compare 9.4). To each animal at least 4 m^2 of green roaming area in a perimeter of 150 m has to be available. Measures have to be taken that a nutrient intake of 170 kg N per ha roaming area and year must not be exceeded. Strongly used areas close to the barn are to be strewn with bark shred or similar and designed in such a way that the strewing, soil material respectively, enriched with nutrients, can be replaced periodically, latest before next recharging of the barn. Plants must grow on the majority of the outside roaming area. Frequent and sufficient resting periods have to be scheduled for the regeneration of vegetation.

Access to the green roaming area has to be provided during the whole vegetation period from 12.00 h noon on. In extreme weather conditions (snow, permanent rain, thunderstorms) access to green roaming area can be restricted in time or totally. The green roaming area has to offer protection from enemies and shade to the animals, so that they will use the roaming area in an equally distributed manner. Thickets will be planted for a natural structure of the whole roaming area. Shading or wind protection nets will provide artificial possibilities of shelter.

Young Hens

The regulations on laying hens as described above apply for young hens respectively as far as in the following paragrafs no other regulations to be met. Additionally the following has to be observed:

Principle

During growth the young animals should learn the natural behaviour which they can conduct in the laying barn, that way avoiding behavioural disorder. During growth robustness should be Bioland Standards April 26th, 2005 page 12/12 developed and a natural immunisation should be achieved. The housing system in the growing barn should be equivalent to the barn of the laying hens.

Barn

In the prime weeks of life rings for chicken are permissible. From the 3rd up to the 12th week of life max. 15 kg life weight and not more than 16 animals per m^2 moving area are allowed to be kept. From the 12th week of life per m^2 of accessible moving area a maximum of 10 animals can be kept in the barn.

In barns with multiple levels a maximum to 24 animals per m^2 barn floor area can be kept from the 12th week of life on. In barns with an integrated exterior climate area in the 12th week of life, at night times, a maximum of 13 young hens per m^2 accessible moving area can be kept in the heated area, provided the exterior climate area is permanently accessible during the light phase. At the earliest from the 6th life week on in addition the exterior climate area (wintergarden) may be taken into account. At least half of the moving area of the barn must be provided as area for scratching purposes. The strewing material has to be at least 5 cm deep and to be kept loose, dry and clean. Daylight with natural intensity is obligatory. If suitable equipment is installed, the application of a lighting program can limit light exposure and period.

Clean and fresh drinking water is always supplied to all animals. The equipement for feeding shall be constructed in a way that all annimals are able to eat at the same time. From the 1st week of life on the animals must be provided with raising possibilities, from the 8th life week on 8 cm per animal, from the 12th week on 12 cm of perching rod per animal is required, whereas 1/3 are to be designed as elevated perching rods. From the 1st week of life on the animals must have available a dust bath and strewing material with sand and coarse-graind limestone material as well as opportunities for shelter and to cover.

Exterior Climate Area and Outdoor Run

From the 10th week of life latest the animals must have access during their activity period to a durable, roofed exterior climate area in the size of at least one quarter of the accessible barn area, depending on the extent of feathering and the climate. The size of the barn openings is at least 2 m per 1000 young hens. Stock sizes of less than 200 young hens or mobile stables are excluded, if a green open air run of at least 2,5 m^2 per animal is available. It shall be possible to divide the green outdoor run into portions, and it must contain equipment for shelter

Poultry for Fattening

The regulations on the keeping of laying hens apply for the keeping of poultry for fattening respectively. Additionally the following applies:

Barn

Locally the single barn buildings with a maximum of 4800 fattening chicken, 5200 guinea-fowls, 4000 female ducks, 3200 male ducks or 2500 geese and turkey hens have to be separated in such a way that infections and/or contamination with parasites are reduced, and a sustainable management of the green roaming area is achieved (for existing barns transitional period until 2010 is possible with the permission of the BIOLAND Association).

The total used area of all fattening poultry barns of one single business may not exceed 1600 m^2 (for existing barns transitional period until 2010 is possible with the permission of an inspection authority, compare 9.4). In the barn per m^2 of accessible roaming area a maximum of 21 kg live weight and not more than 10 animals can be kept, whichever case applies first. Roaming area is defined as the floor area of the barn available to the animals. In mobile barns a maximum of 30 kg live weight and not more than 16 animals can be kept per m^2. For guinea-fowls a minimum of 20 cm perching rod per animal is mandatory. For fattening chicken and turkey hens perching rods in respect to their size and age are to be offered.

Exterior Climate Area

For fattening chicken and turkey hens an exterior climate area or a durable roaming area is mandatory in addition to the interior barn area. The size has to be at least one third of the minimum barn area. Exempt from this are stock sizes of less than 100 animals and mobile barns. The exterior climate area can be taken into the calculation of the animal density.

Greened Roaming Area

A greened roaming area is mandatory (for existing barns for fattening chicken with respect to the afforded greened roaming area, for other fattening poultry with respect to the size of the greened roaming area respectively, transitional period until 2010 is possible with the permission of the BIOLAND Association, compare 9.4). To every animal the following minimum greened roaming area per animal has to be afforded:

- fattening chicken and guinea-fowls 4,0 m^2
- ducks 4,5 m^2
- turkey hens 10 m^2
- geese 15 m^2
- fattening poultry in mobile barns 2,5 m^2.

If the climatic conditions and the physiological conditions of the animals allow for this, access to a greened roaming area has always to be afforded, nevertheless if possible at least during one third of the lifetime. Restrictions may result from the physiology by the age and by the feathering of the animals and from the climate.

Water Surfaces

Water fowl has to be afforded access at any time to running streams, ponds or lakes (only if hygienic conditions and water protection acts permit it) or to a durable water surface that is replaced regularly by fresh water.

Keeping Horses

Whenever soil conditions allow, horses have to be afforded grazing or exterior roaming. For them being kept in barns it has to be in the form of boxes or non-penned stables with access to open air roaming area, if possible. As far as possible, the animals are to be kept in groups.

Keeping of Fallow-Deer and Red Deer

For fallow-deer and red deer keeping on pasture is mandatory all year round. The minimum preserve size for fallow-deer is 3 ha, for red deer 5 ha. In the pens there must be hiding places for calves. The pens must provide shelter against climate conditions, preferably by means of natural hedges and trees. Red deer pens must have a slough additionally. The minimum pack size of deer comprises of 5 animals (1 stag, 4 females). Per ha preserve area the stock density is 7 PED or 3.5 PER respectively. One production unit fallow-deer (PED) comprises of 1 adult animal, 1 calf, 1 one-year-old, one stag proportionately; one production unit red deer (PER) comprises equally.

Keeping of Rabbits

The following regulations apply for keeping systems of more than 3 animals for breeding purposes or more than 20 animals for fattening respectively.

General

Barn and roaming area allow for the kind of behaviour peculiar to the species. Keeping in groups is mandatory except during the nursing time of females. The maximum group size is limited to 40 animals for fattening purpose or 10 animals for breeding purposes respectively.

Barn

Daylight inside the barn is mandatory. The height of the barn must be at least 60 cm. The area of movement can spread on multiple levels. It ought to contain different surface features. Possibilities for retreat and resting areas must be available for all animals. Each nursing female needs its own nest to litter.

Roaming area

A durable roaming area is prescribed except during times of nursing and in the case of mobile barns. If a greened open air run is provided, areas for rotation and resting periods for the vegetation is mandatory for the run to remain mostly greened and to keep the load of parasites small. In the case of mobile barns there have to be greened open air runs.

Dealing with Animals

General, Dealing with animals must take into consideration the needs of the species and the feelings of the animals.

Measures in the Business

Care of hair, skin and hooves is to be carried out at regular intervals. As far as this is possible in the system used in keeping, no removal of horns should be carried out in the case of ruminants.

Not permissible are:

- removal of horns by means of cautery sticks
- cropping of tails in the case of cattle or pigs
- prophylactic shortening of pigs' teeth
- insertion of nose rings and nose clamps to-prevent pigs from grovelling
- disfigurement of poultry by shortening of the beaks, cropping of comps and wings.

Animals must not be subjected to further surgical interference systematically. Keeping a laying pause is possible for laying hens. During this resting period the free access to water and fodder may not be restricted. The day light may be limited to 5 hours per day. As far as possible in each flock at least 1 cock per 100 hens should be kept from the beginning of the rearing time.

Chapter 11

Genetic Engineering and Traditional Breeding Methods: A Technical Perspective

The Soil Association's rejection of the use of genetic engineering (GE) in agriculture as simply having "no place in organic food and farming" (Living Earth, Jan. '97), is justifiable purely as a matter of principle. GE represents an extension of intensive, industrial agriculture and therefore reinforces environmentally damaging, non-sustainable husbandry. Evidence already exists which demonstrates that the claims that GE crops will result in less dependence on agrochemicals are, in the medium to long term, unfounded.

The greatest claim of those who endorse the use of GE in agriculture, is that it is a safe, more precise and natural extension of traditional cross breeding methods for generating novel varieties of crops and farm animals. It is said that this new technology simply gives nature a helping hand with something that would happen anyway. The aim of this article is to assess GE in agriculture from a technical and basic genetics viewpoint focusing in particular on plants and animals. We will see that technically speaking, the use of GE in agriculture is a crude and imprecise technology which bears no resemblance to traditional breeding methods for producing new varieties of crops and farm animals. Given this imprecision, the outcomes of using GE in food production both in terms of potential ill health and negative environmental impact, are far from certain. There would therefore also appear to be good scientific grounds for questioning the validity of using GE in agriculture especially when there are safe alternatives available.

The Fundamentals-Genes and Genetics

Genes are discrete units of DNA. They are the blueprints which carry the information for the tens of thousands of proteins which act as the building blocks of all the structures and functions (biochemistry) that constitute the body of any organism from bacteria to humans. DNA can be likened to a long string of pearls where each pearl, representing a gene, occupies it's own special place in the "necklace" which is vital for it's correct function. Genetics, the study of genes, has two basic components. Firstly, there is the information content of each gene; that is, what gene carries the blueprint for which protein. Secondly, genetics has taught us that the activity or expression of each gene is extremely tightly controlled or regulated. Put simply, each gene has it's own set of sophisticated on-off switches to drive it's expression ensuring that the correct protein and therefore appropriate structure and function, is present in the right place, time and quantity in the body.

Just as all forms of life are interdependent upon each other for survival and growth, no gene works in isolation from all other genes. *The latest discoveries tell us that genes are arranged along the DNA in groups or "families". The function of a given gene in a group is dependent on all the other genes that are present within the same family. Furthermore, the genetic activity in one family of genes can effect the function of genes in other groups of genes. It is also clear that genes and the proteins that they give rise to, have co-evolved together to form an extremely intricate, interconnected network of finely balanced functions the complexities of which we are only just beginning to understand and appreciate.*

Such tight control of gene activity means you will never find liver functions in your brain or leaf specific processes in the fruit and vice versa! In addition, Nature has also evolved mechanisms whereby cross breeding can only take place between very closely related species. With traditional breeding methods, different variations of the same genes in their natural context (within the necklace of pearls) are exchanged. This preserves tight control and complex interrelationships between genetic and protein functions that are vital for integrity of life as a whole.

GE: A Natural Extension of Traditional Breeding Methods?

In order to assess the validity of the claim that GE represents a natural extension of traditional breeding methods, it is important to know how GE ("transgenic") plants and animals are produced.

GE Plants

As an example, let us see how the herbicide resistant, GE soya was generated. The objective here was to introduce into the soya plants a gene from a common soil bacterium which would allow it to survive when sprayed with the herbicide Roundup. Clearly you cannot "cross" a bacterium with a plant.

Therefore, the first step was to grow cells from soya bean plants on plastic dishes in the laboratory. Now, in order to allow the bacterial gene to be able to work once introduced into it's new plant host, it had to be linked to a genetic switch combining parts from a cauliflower virus and petunias. (As we discussed above, the bacterial gene's own switch will only work in the bacteria from which it came). This combination of cauliflower virus, petunia and bacterial DNA was then introduced into the soya bean cells growing on the dishes in the laboratory using a procedure known as "biolistics" which employs a device called a "gene gun". In this technique, tiny spheres of gold or tungsten are coated with the DNA one wishes to introduce into the plant cells. These DNA-coated metal particles are then shot at the plant cells using the gene gun at high speed. As a result some of these metal beads enter inside the plant cells carrying the new DNA with them.

Unfortunately from the point of view of the plant biotechnologist, the efficiency with which the new DNA is taken up by the soya bean cells on the dish is very low. Most of the cells don't take it up at all. So the key is to find those few cells among the many millions on the dish which have taken up the DNA. This is done by using another genetic trick. The introduction of the bacterial gene into the soya bean cells for herbicide resistance, was accompanied by a second gene which confers resistance to an antibiotic (called kanamycin). The soya bean cells were then treated with the antibiotic. The few cells which had taken up the herbicide resistance: antibiotic resistance "marker" gene combination survived and flourished whereas the majority of the cells which had not taken up these genes were simply killed by the antibiotic.

Finally, by changing the conditions under which the soya bean cells are grown, the cells clump together to form what is called a callus which in turn starts to put down roots and sprout green shoots. These little "seedlings" are then potted so as to grow into fully mature plants which will carry in all their cells (including those for reproduction; i.e. pollen etc.) the new bacterial gene. The plant which then displays the best agronomic performance, in this case resistance to herbicide, is then selected for further development (crossing to form new hybrids etc.).

GE Animals

The generation of transgenic animals is a somewhat simpler, but no less artificial procedure. Fertilised eggs are first removed from the animal of choice. These eggs are then injected with the genes one wishes to engineer into the animal. The DNA injected eggs are then returned to the womb of a surrogate mother where they complete their development and are born in due course. Therefore, in marked contrast to traditional breeding methods, all transgenic plants and animals start life as individual or groups of cells growing on a plastic dish in a laboratory.

GE: A No Holds Barred Technology

It is evident from the procedure we just described that with GE there are no holds barred. GE allows the isolation, cutting, joining and transfer of single or multiple genes between totally unrelated organisms circumventing natural species barriers. As a result combinations of genes are produced that would never occur naturally. Transgenic crops containing genes from viruses, bacteria, animals as well as from unrelated plants have been generated.

In the case of the herbicide resistant soya beans, the final outcome was the combination of genetic material from four totally unrelated organisms; a cauliflower virus, petunia, bacteria and soya. Furthermore, again as we saw in the case of the GE soya beans, the newly introduced gene units are composed of artificial combinations of genetic material. Another example which illustrates the extreme combinations of genetic material that can be produced, is the introduction of the "anti-freeze" gene from an arctic fish (the sea flounder) into tomatoes, strawberries and potatoes in the hope of producing resistance to frost.

As with the bacterial gene in the soya beans, the fish anti-freeze gene is joined to the cauliflower virus genetic switch to allow it to turn on and work in it's new host. (The fish genetic switch naturally only works in the fish). All this is in turn coupled to an antibiotic resistance marker gene to allow selection of the newly transformed plants.

GE Disrupts Host Gene Functions and Possesses Inherent Unpredictability

Clearly GE represents a great technological advance. However, as we have already discussed, genes have evolved to exist and work in families. Therefore, the claim that the reductionist approach of GE which moves one or a few genes between unrelated organisms, is a precise technology is highly questionable.

Furthermore, the generation of transgenic plants and animals is currently an imperfect technique. Once injected into the cells of the organism, the introduced gene is randomly incorporated ("spliced") into the DNA of it's new plant or animal host.

In fact, the manner in which GE animals and plants are produced, always selects for the splicing of the foreign gene into regions of the host DNA where other natural genes are trying to work. Given the interdependence of gene function within any grouping of genes, this random splicing of the foreign gene into the host DNA will always result in a disruption in the normal genetic order in the "string of pearls". Therefore, GE of animals and especially plants, always results in a loss, to a lesser or greater degree, of the tight genetic control and balanced functioning which is retained through conventional cross breeding. With GE, host genes can be silenced (inactivated) or inappropriately switched on resulting in either a deficiency in a given protein(s) or the presence of the wrong protein(s) in the wrong place or in the wrong quantity or all these combined.

In addition, it is also assumed that the introduced gene and the protein that it makes, will behave in exactly the same way in it's new host as it does in it's native environment which frequently will not be the case. As discussed above, gene and protein functions have evolved over millions of years to work together in any given organism. The anti-freeze gene/protein in the arctic sea flounder has evolved to work together with the other genes/proteins in this fish. It is purely an assumption that it will work in exactly the same way with no unwanted side effects in it's new hosts where it will now be surrounded by plant proteins!

These effects combine to produce a totally unpredictable disturbance in host genetic function as well as in that of the introduced gene. The resulting disturbance in biochemical function can unexpectedly produce novel toxins, allergens and reduced nutritional value.

Conclusion: GE and Traditional Breeding Methods Are Worlds Apart

The proponents of the use of GE in agriculture argue that mankind has been selecting and manipulating plant and animal food stocks for millennia and that this new technology is simply the next stage in this process. However, we have seen:

- Technically speaking, GE and traditional breeding methods bear no resemblance to each other.
- GE plants and animals start out life in a laboratory culture dish.

- GE employs totally artificial units of genetic material which are introduced into plant and animal cells using chemical, mechanical or bacterial methods.
- GE always results in disruptions to the natural order of genes within the host DNA.
- GE also brings about combinations of genes that would never occur naturally.

Clearly these procedures are worlds apart when compared to cross fertilisation between closely related species. The totally artificial nature of GE does not automatically make it dangerous. It is the imprecision in the manner by which genes are combined and the unpredictability in how the introduced gene will interact within it's new environment which results in uncertainty. The balanced gene functions that have evolved together and which are preserved with traditional methods, are lost with GE. Therefore, from the standpoint of the fundamental principles of genetics and the limitations in the technology, GE is neither more precise nor a natural extension of traditional cross breeding methods. If anything the opposite would appear to be true. Therefore GE foods possess new and unique safety considerations both in terms of health and to the environment.

The availability of safe, sustainable, natural methods of breeding and husbandry utilising the many thousands of different varieties of a any given food crop, makes the risks associated with GE foods simply not worth taking. These risks are even less acceptable when one takes into account the fact that once released into the environment, genetic mistakes/pollution cannot be recalled, cleaned up or allowed to decay like agrochemicals or a BSE epidemic, but will be passed on to all future generations indefinitely.

Breeding and Supply

Many animal models of human disease are chosen because they may be readily bred in captivity, enabling many generations to be evaluated in a relatively short period of time. Genetically altered (GA) animals are also important, often as models of specific human disease. The links above provide information on breeding of the most commonly used vertebrate species, including GA mice. Although invertebrates such as the nematode worm *Caenorhabditis elegans* and the fruit fly *Drosophila melanogaster* are important to biomedical research, a discussion of breeding techniques of invertebrates is beyond the scope of this Information Portal.

Many laboratory animals are obtained from commercial breeders which are able to supply a wide range of well-defined, high quality animal models representing most of the commonly used laboratory species. In-house breeding is largely confined to the stocks and strains of animal models that are not commercially available, the production of most transgenic animals that now comprise a significant and increasing proportion of those used in biomedical research, and/or to the maintenance of colonies of animals associated with studies of reproduction and genetics.

Breeding methods which are suitable for the production of animals from in-house colonies are outlined in Festing and Peters (1999). Details of genetic considerations in the maintenance of inbred, outbred and mutant rodent stocks are discussed in Festing (1999), and methods of refinement and reduction in the production of GA mice are set out in Robinson et al. (2003). The report of the RSPCA Resource Sharing Working Group gives advice on archiving and sharing of GA mice lines. Although techniques will differ from species to species, some basic principles apply to the breeding of all laboratory animals:

Reproduction and reproductive performance may be influenced by a variety of extrinsic factors and consequently the physical environment is important for maintaining the health and welfare of breeding animals, and for enhancing their breeding performance. Optimal ranges for environmental conditions, as well as minimum standards of husbandry and care, are set out in the Code of Practice for the Housing and Care of Animals in Designated Breeding and Supplying Establishments.

As breeding animals are typically maintained for longer periods than animals used on scientific procedures, particular attention is needed to ensure that the environment provides for the animals behavioural as well as physiological needs. The process of breeding laboratory animals can involve the thwarting of natural behaviours, e.g. laboratory animals may be weaned and separated from their dam at a time which rarely coincides with the time they would have dispersed naturally, and this can be stressful for both the juvenile animals and the parent. It is also important to bear in mind that the needs of infants and juveniles may be different from those of adults.

In situations where breeding stock and/or post-wean stock are housed in groups, individuals that may be disadvantaged in the social hierarchy, e.g. subordinates and females that have recently given birth, may be vulnerable to social stresses. Extra care should be taken to prevent and monitor aggression and to separate individuals if necessary. Reproduction is an energy intensive activity, so animals in poor condition should not be used for breeding.

Breeding animals should be provided an appropriate diet which should be formulated to satisfy their nutrient and energy requirements, particularly for the demands of pregnancy and lactation, and be palatable and free from chemical and biological contamination.

Health status has a significant influence on animal welfare and the use of healthy animals is regarded as a prerequisite both for good welfare and for good science. Although animals may be bred for use as specific disease models, intercurrent disease within the population may call into question the validity of information obtained from scientific procedures and make interpretation of results difficult or even impossible. It is important, therefore, to have a system for monitoring animal health and to have plans to maintain and deal with potential problems, such as disease outbreaks. FELASA has published some recommendations, relating predominantly to the monitoring of microbiological status, for laboratory animal health screening programmes.

Research animal models will invariably be chosen on the basis of particular genetic and/or phenotypic characteristics. When breeding animals for research, therefore, appropriate resources should be directed towards monitoring these essential traits and in these and other ways, maintaining the genetic and phenotypic integrity of the colonies.

Regular handling will accustom young animals to human contact, making future handling and restraint for procedures less stressful for the animal and easier for the handler. For the larger species, attention should also be paid to appropriate habituation and training so as to prepare the animals for their life on study.

Matching Supply and Demand

It is essential to try and ensure that the supply of animals does not exceed the research requirement in order to avoid the generation of an unnecessary surplus and the potential wastage of animals lives. This requires good communication and coordination between breeders, suppliers and users (which is more feasible with in-house breeding), and also within and between multidisciplinary research teams. It also requires good production planning.

A LASA task force has produced a report on the production and disposition of laboratory rodents surplus to the requirements for scientific procedures, which makes recommendations for minimising avoidable excess. The MRC has issued codes of practice for the supply of rodents and supply and use of aquatic species in research.

In most cases, especially with small rodents, supply and demand cannot be balanced exactly. For example, there is often a marked difference in the requirement for either males or females. This generates an inevitable biological surplus to which must be added a managed surplus arising out of variable management practices and user requirements. Where surpluses do occur it is important to review the causes and take appropriate action to try and prevent the situation recurring.

Rehoming may be an option for species that are kept as companion animals if any surplus cannot be redirected for use in essential procedures. Clearly individual animals should only be rehomed where it is in their best interests, where their welfare can be assured, where adequate resources are available for their long-term care, and with the proviso that they are not suffering or likely to suffer any adverse effects from their use in research.

Whenever animals are euthanased every effort should be made to bank and use their tissues and blood if this will avoid the unnecessary killing of another animal.

Information on the production, supply and usage of animals should be provided to, and considered within, each establishments ethical review process to ensure the fullest use is made of individual animals and that numbers bred are reduced to a practicable minimum.

Hybridization and Genetic Engineering

Animals of domestic origin as well as feral animals sometimes produce fertile hybrids with native, wild animals which leads to genetic pollution in the naturally evolved wild gene pools, many times threatening rare species with extinction. Cases include the mallard duck, wild boar, the rock dove or pigeon, the Red Junglefowl (Gallus gallus) (ancestor of all chickens), and carp. Another example is the dingo, itself an early feral dog, which hybridized with dogs of European origin. Genetic pollution is a serious issue: Living organisms can also be defined as pollutants, when a non-indigenous species (plant or animal) enters a habitat and modifies the existing equilibrium among the organisms of the affected ecosystem (sea, lake, river). Non-indigenous, including transgenic species (GMOs), may bring about a particular version of pollution in the vegetal kingdom: So-called genetic pollution. This term refers to the uncontrolled diffusion of genes (or transgenes) into genomes of plants of the same type or even unrelated species where such genes are not present in nature. For example, a grass modified to resist herbicides could pollinate conventional grass many miles away, creating weeds immune to the most widely used weed-killer, with obvious consequences for crops. Genetic pollution is at the basis of the debate on the use of GMOs in agriculture.

A Genetically Modified Organism (GMO) is an organism whose genetic material has been altered using the genetic engineering techniques generally known as recombinant DNA technology. Genetic Engineering today has become another serious and alarming cause of genetic pollution because artificially created and genetically engineered animals in laboratories, which could never have evolved in nature even with conventional hybridization, can live and breed on their own and, what is even more alarming, interbreed with naturally evolved wild varieties.

Common Sub-divisions of Animal Husbandry

Historically, animal husbandry is broken up into animal subdivisions, agriculturists tending to specialize on one or two types of animals. Below is a list and brief description of some of the major subdivisions:

- Swineherd: A person who cares for hogs and pigs (older English term: Swine).
- Shepherd: Usually a person who cares for sheep. However, in previous years, it was common to have herds which were made up of both sheep and goats, in which case the term shepherd was still employed.
- Goatherd: Someone who cares primarily for goats.
- *Cowherd/Cowboys: Refers to those who raise, milk, and slaughter cattle. In more modern times, the cowboys or* vaqueros *of North and South America ride horses and participate in cattle drives to watch over cows and bulls raised primarily for food.*
- Herder: A generic term for anyone who tends to a herd of animals not listed above, such as Camels, yaks, and in Latin America, llamas and alpacas.
- Equestrian: A person who breeds and takes care of horses.
- Breeders: Generic term for anyone who specializes in the breeding of agricultural animals. More often, breeders are trained in specialized techniques, such as artificial insemination and embryo transfer, and are therefore independent hires of the usual farm staff.

An Overview of Animal Breeding

The face of animal breeding has changed significantly over the past decades. Animal breeding used to be in the hands of a few distinguished 'breeders', individuals who seems to have specific arts and skills to 'breed good livestock'. Nowadays, animal breeding is much dominated by science and technology. In some livestock species, animal breeding is in the hands of large companies, and the role of individual breeders seems to have decreased. There are several reasons for this change. Firstly, the breeding industry has taken up scientific principles. Looking was replaced by measuring, and an intuition was partly replaced by calculations and scientific prediction. Other major developments were caused by the introduction of biotechnology. These are roughly the reproductive technologies, and the molecular genetic technology. Not all of this is new. Artificial insemination was introduced in the fifties in cattle.

No doubt that the technology had a major impact on rates on genetic improvement in dairy cattle, and just as important, on the structure of animal breeding programs. Nowadays, technologies like ovum pick up, in vitro fertilization, embryo transfer, cloning of individuals, cloning of genes, and selection with the use of DNA markers are all on the ground. Some of the technologies are already applied, others are further developed, or waiting for application. Finally, the rapid development of computer and information technology has greatly influenced data collection and genetic evaluation procedures in livestock populations, now allowing comparison of breeding values across herds, breeds or countries. The introduction of breeding methods typically needs to find the right balance between what is possible from a technological point of view and what is accepted by the decision makers and users within the socio-economic context of a production system. Ultimately it is the consumer who decides which technology is desirable or not.

In most western societies, consumers are increasingly aware of health, environmental and animal welfare issues. Food safety and methods food production are part of their buying behaviour. However, price and production efficiency remain to be major contributors to sustainability of a livestock industry. Successful animal breeding programs need to find the right dose of technology that helps them to be competitive.

Breeding Strategies

Reproductive rate of breeding animals and uncertainty about true genetic merit of breeding animals make up the most important limiting factors in a breeding program. How many and which animals should be selected is determined by these factors. Investments in breeding programs are therefore often related to trait measurement and genetic evaluation, and to technology to increase reproductive rates.

Measurement Effort and Genetic Evaluation

The benefit of abundant and good measurement is that we may better be able to identify the genetically superior animals. This leads to more accurate selection and more genetic improvement.

Phenotypic measurements are turned into *Estimated Breeding value's (EBV's).* Estimation of breeding value based on an animal's phenotype alone can already be quite accurate for high heritable traits. However, animals need to be compared across herds, and genetic and environmental influences have to be disentangled. To achieve this, more sophisticated statistical methods are used, leading to *Best Linear Unbiased Prediction* (BLUP) of breeding values. Besides allowing across herd comparisons, BLUP also uses all available information about an animals' breeding value, including data on related animals. Selection accuracy is strongly dependent on the degree of data recording, which requires a range of considerations related to cost and infrastructure.

In data recording, individual performances need to be related to animal identification. If BLUP is used to generate EBV's also an animal's pedigree needs to be known (in principle, for each animal only sire and dam). If pedigree is not recorded, breeding value can be assessed on own performance only, and is limited to sexes, which express the traits of interest. BLUP relies good structure of data (use of breeding animals across herds) and proper pedigree recording. If these prerequisites are in place, investment in BLUP methodology is usually highly cost efficient. Molecular genetic technology has rapidly developed in the past 2 decades. Genes have been found coding for factorial traits (such as many diseases). Many production traits are *quantitative traits* and a likely genetic model is here that genetic differences between animals are due to many genes. However, DNA technology has also provided genetic markers. Certain genetic markers can improve estimation of an animal's genetic potential as they are associated with regions that account for genetic variation. Genotyping animals for marker genotypes is therefore an investment with the aim to better assess true genetic merit of animals.

Reproductive Technology

Most of the main factors that determine genetic gain are directly influenced by the reproductive rate of the breeding animals. A higher reproductive rate leads to the need for a decreased number of breeding animals, therefore increasing the intensity of selection of these animals. If reproductive technology is possible, for example AI, the benefit could be expressed in terms of increased genetic rate of improvement, which in turn has a dollar component attached to it. More offspring per breeding animal allow also more accurate estimation of breeding value. Reproductive technology allows the intensive use of superior breeding stock.

An obvious consequence is possibly that the most popular breeding animals are overused, and the population could encounter inbreeding problems. Typically, as new technologies in animal breeding allow faster genetic change, long term issues such as inbreeding and maintenance of genetic variation become important. For that reason, selection tools in animals breeding have become somewhat more sophisticated in recent years. The impact of reproductive technologies on rates of genetic improvement and inbreeding will be discussed. Besides a direct effect on rate of genetic improvement, another important consequence from increasing reproductive rates is to disseminate superior genetic stock quickly. The influence of a superior breeding animal would be much higher if thousands of off spring could be born, rather than if the superiority is passed on through the production of sonsvia natural mating.

Another example is that of cloning. Cloning is not extremely important for increasing rate of genetic progress, but it could have a large impact by allowing many copies of the best individual to perform in commercial herds. As reproductive rates are basically multiplying factors in a breeding structure, any improvement in reproduction will justify higher investment in improvement of the best breeding stock.

Selection and Mating

The decision about which animals should be selected as parents for the next generation is mainly based on *assessment of breeding value* of individual animals. *Genetic evaluation* is central to animal improvement schemes. Selecting animals based on estimated breeding value maximizes the response to selection that can be achieved. However, there is one other criterion that is relevant when deciding which animals should have offspring.

This criteria is *common ancestry* of all selected parents. The coancestry of selected parents should stay below certain limits, since it is directly related to the build up of inbreeding. Coancestry among selected parents is determined by the average relationship among the selected parents as well as the number of parents selected. In this course we will more explicitly discuss selection strategies that maintain low levels of inbreeding. Decisions about which animals need to be mated are often seen in relation to dominance effects. Utilizing dominance variation is often not of primary importance for improvement of purebreds, but it can have more impact if breeding animals are selected from different breeds or lines, as heterotic effects between breeds can be utilized.

When multiple traits are involved in the breeding objective, assortative mating could be useful, matching qualities in different parents for different. There is a good possibility that in the near future, planned mating will gain in importance, when effects of specific genotypes will be better understood. One could envisage certain genotypes with high growing potential to be combined with specific genes that have major effect on meat quality. Overview of animal breeding programmes avoid inbreeding in direct offspring as well as the rate of inbreeding in the population.

However, the rate of inbreeding depends mainly on population size and number of parents selected. Methodology to optimize selection and mating decisions related to inbreeding will be discussed.

Structure of Breeding Programs

Most of the key decision factors mentioned earlier are related to the rate of genetic change that can be made. However, this could be genetic change in a small fraction of the national population (in nucleus or 'elite breeders'). Genetic superiority should be transferred as soon as possible to most of the commercial farms.

The structure of a breeding program is therefore relevant for two aspects of an improvement scheme:

Introduction to Marker Assisted Selection

Over the last two decades most livestock industries have successfully developed EBV's to allow identification of the best breeding animals. EBV's are best calculated using BLUP, meaning that they are based on pedigree and performance information of several traits from the individual animal and its relatives. BLUP EBV's are the most accurate criteria to identify genetically superior animals based on phenotypic performance recording. Although the idea of genetic selection is to improve the genes in our breeding animals, we actually never really observe those genes. Selection is based on the final effect of all genes working together, resulting in the performance traits that we observe on production animals. This strategy makes sense, since we select based on what we actually want to improve. However, animal performance is not only affected by genes, but also by other factors that we do not control. Selection for the best genes based on animal performance alone, can never reach perfect 100% accuracy.

A large progeny test comes close such a figure of perfect selection, but this is expensive for some traits (e.g. for traits related to meat quality), and we have to wait several years before the benefits from a progeny test have an effect. Efficient breeding programs are characterised by selecting animals at a young age, leading to a short generation intervals and faster genetic improvement per year. For selecting at younger ages, knowledge about the existence of potentially very good genes could be very helpful.

Quantitative genetics uses phenotypic information to help identify animals with good genes. *Extension to use information from molecular* genetics techniques aim to locate and exploit gene loci which have a major effect on quantitative traits (hence QTL-Quantitative Trait Loci). The idea behind marker assisted selection is that there may be genes with significant effects that may be targeted specifically in selection. Most traits of economic importance are quantitative traits that most likely are controlled by a fairly large number of genes. However, some of these genes might have a larger effect. Such genes can be called major genes located at QTL.

In practice, we rarely know the genotype at actual QTL, as the exact gene location (mutation) is often unknown. Currently there are few examples where QTL effects can be directly determined, but knowledge in this area is rapidly developing. Most QTL known today can only be targeted by *genetic markers.* Genetic markers are "landmarks' at the genome that can be chosen for their proximity to QTL. We cannot actually observe inheritance at the QTL itself, but we observe inheritance at the marker, which is close to the QTL. When making selection decisions based on marker genotypes, it is important to know what information can be inferred from the marker genotypes.

We can identify the marker genotype (Mm) but not the QTL genotype (Qq). The last is really what we want to know because of its effect on economically important traits. Let the Q allele have a positive effect, therefore being the preferred allele. In the example, the M marker allele is linked to the Q in the sire. Progeny that receive the M allele from the sire, have a high chance of having also received the Q allele, and are therefore the preferred candidates in selection.

Mixed Models in Animal Breeding: Where to Now?

Over the past 60 years, mixed models have underpinned huge gains in plant and animal production through genetic improvement. Charles Henderson (1912-1989) established mixed models for estimating breeding values (BLUP) using the popularly called Henderson's Mixed Model and provided early methods (Henderson's Methods I, II and III) for estimating variance parameters. Robin Thompson then published the widely acclaimed REML method for variance component estimation in 1971. These two innovators, along with the development of computing power, have spawned national and international breeding programs in almost all animal species used for human food and fibre.

Our ability to generate data is outstripping our ability to analyse data and this will lead to mixed models playing new roles in genetic estimation. The focus is changing from simply describing the relationship between variables through a correlation, to modelling the relationship based on knowledge of the Genome.

Selective breeding goes back at least to Jacob (1800 BC, Genesis 30) who selected the fitter rams for his own flock. Traditional breeding has largely relied on visual assessment with many such classers having considerable skill in recognising genetic potential with respect to their objective, whether breeding war horses, dogs or pigeons. What characterises modern breeding though is the extensive use of objective measurement and adjustment for environmental effects.

The digital age has seen a rapid increase in the number of traits included in a breeding objective or selection criterion, as well as use of data on relatives to improve the separation of genetic from environmental differences. Charles Henderson (1912-1989) *et al.* (1949, 1959) developed and popularised the mixed model equations which underpin the BLUP estimation of breeding values. His development of these equations included use of the additive genetic relationship matrix, showing how it accommodates selection as well as their primary role of adjusting for nuisance environmental effect

However, the mixed model equations used for evaluation assume knowledge of variance parameters. Henderson (1953) defined the main methods used to estimate these until Robin Thompson (Patterson and Thompson 1971) presented the Residual Maximum Likelihood (REML) method. Karin Meyer and Dorothy Robinson produced software to implement REML methods (in animal breeding and However analysis was difficult until Robin presented the Average Information method underpinning ASReml (1997, 2002, 2006, 2009) which become generally available in 1997.

The promise of the genomic revolution is that we may be able to select directly for specific combinations of genes based on reading an individual's genetic code and having good information on the phenotypic and pleiotropic effects of genes/alleles.

Mixed Model Equations and Blup

The linear mixed model is written as where X is the design matrix for fixed effects, τ, Z is the design matrix for random effects, u, y is the vector of phenotypic measurements and e is the vector of model residuals. The mixed model equations (MME) are conveniently represented in matrix form by where var. Given R and G, the solution for the fixed effects given by the mixed model equations is the same as given by solving where. The solutions for the random effects are the Best Linear Unbiased Predictors of those effects and as such are ideal for selecting breeding stock.

The power of this system lies in the structure that can be incorporated into X, Z, R and G. It is is not unusual for u to include sub-vectors for various traits and various 'strata' such as direct genetic, maternal genetic, maternal environment, dominance and nuisance blocking effects. This can lead to a fairly complex structure to G involving relationship matrices and variance matrices of various sorts. The main advantage of the mixed model equations is that the left hand side matrix is typically fairly sparse so that large systems of equations can be solved quite efficiently. This arises because matrices X and Z, and inverses of R and G are typically sparse.

Residual Maximum Likelihood

Without going into the detail, suffice to say that if we assume u and e (and therefore y) are normally distributed (given R and G), we obtain an expression involving y, X, Z, R and G which is called the likelihood. This expression can be partitioned into two parts; one providing information on τ conditional on G and R leading to the mixed model equations, the other providing information on R and G conditional on τ. Residual Maximum Likelihood seeks to find the parameter values for R and G that are most likely because they maximise this second part (rather than the whole likelihood). This maximisation exercise though was not trivial when R and G involved more than a few parameters and the problem was large.

Consequently, REML estimation was restricted in application to small problems or well structured standard animal breeding models until Thompson presented the Average Information procedure which is also centred around the mixed model equations. The implementation in ASReml exploits the sparsity of the mixed model equations though judicious ordering of the equations, avoiding the need to obtain the complete inverse of the left hand side matrix. Now REML can be applied to large problems (with several hundred variance parameters).

Where to Now?

One thing programming has taught me is that no matter how big you allow, someone will want bigger. While computing technology has helped with the more traits, more records issue of modern animal breeding based on BLUP technology, we are now faced with genome level data of a higher magnitude and methodologies which do not have the statistical and mathematical rigor that supports conventional quantitative genetics.

Three problem areas come to mind. The first is the well established variance estimation problem (Hill and Thompson 1978) that when estimating a variance matrix, the probability that the maximum value of the REML likelihood occurs outside the imposed parameter space increases with the matrix size.

The second is the application of mixed models to genomic data. The third is how to effectively combine specific genomic data into the BLUP evaluation process.

Structured Variance models. The more traits involved in a REML analysis, the more likely there will be difficulties with the estimation of all the variances and co-variances involved. ASReml will estimate a negative definite matrix if permitted, or attempt to estimate a positive definite matrix which is almost singular. But this raises the issue of whether a reduced parameterization within the parameter space will be preferable. It is not uncommon to find that a matrix can be reduced by use of principal components to a more parsimonious form. That is, the first 1, 2 or 3 principal components will contain the big bulk of the information contained in the matrix.

The remaining variation is noise and is often associated with negative eigen values. Therefore it makes sense to estimate the matrix based on some underlying structure. Three structures are common in ASReml. For variates that have no intrinsic ordering, the principal component/factor analytic models allow more parsimonious modelling. For measurements repeated at irregular intervals, the random regression models are often applied but these may produce unreasonable estimates at the ends of the time range.

For regular repeated traits, for example weights at successive ages, the expected structure is an autoregressive one for which the Antedependence (Generalised auto regressive) models apply. Jaffrzic et al. (2002) has extended the Antedependence model to a Structured antedependence where a model is imposed on the regression and innovation parameters.

Meyer and Kirkpatrick (2009) have investigated a reduced parameterization based on assuming common eigen vectors across strata which is another proposal within this framework. To my mind, this leads to a general area of writing models for the variance parameters, and is the next logical step when it comes to fitting models with hundreds of variance parameters.

The question will always be whether a reduced parameterization has adequately captured the real variation without imposing a structure unsupported by the data. Mixed models for genomic data.

There is a huge literature on analysing the huge amount of genomic data that is being presented and little consensus on the best approach. One issue is the diversity of kinds of data available and the other is the sheer volume of data and the knowledge that meaningful/useful variation is present in only a small proportion of it.

The issue here is then to separate signal from noise. I believe mixed models could have a bigger role here because signal will represent a covariance (or inflated variance) over the noise (base variance). Mixed models have been successfully used to adjust for spatial variation in genomic slides. They have been used to locate QTL in back-cross/F2 experiments and in association studies where there are often more 'markers' than experimental units.

Thomson et al. (2009) use mixed models as part of their procedure to combined cattle and sheep genomic data to look for differentially expressed genes. The new outlier method in ASReml 3 may help in this regard. Incorporating genomic markers in BLUP evaluation. Scientists are an optimistic group when it comes to incorporating genetic markers into BLUP evaluation. I suspect there is a lot of detailed work required before this becomes standard procedure across the industries.

Discussion

Linear Mixed Models have underpinned a revolution in livestock breeding in the last 50 years and despite the huge investment in genomic research and Bayesian methods, there remains a continuing major role for them in the foreseeable future. However, the general model needs adaption for the specifics of each particular species and application. By this I mean, identification of the principle sources of variation, whether they should be accommodated as fixed or random effects, appropriate variance structures and extending the analyses to larger populations and with more traits.

While a bivariate analysis is now readily performed, larger multivariate analyses for the estimation of positive definite variance matrices are often difficult requiring use of structured matrices and raising the issue of whether the structure is adequate. There will undoubtedly be further developments in this area.

The literature on analysis of genomic data reports a wide range of methods as people have hurried to analyse their large amounts of newly acquired data. Some of these analyses have demonstrated the utility of mixed models in this area, but have also shown up limitations due to the amount and structure of the new data. This also will need more attention.

Bibliography

Adams, Carol J.: *Animals and Women: Feminist Theoretical Explorations.* Durham, NC: Duke University Press, 1995.

Aruna T. Kumar *: Handbook of Animal Husbandry*, Indian Council of Agricultural Research, 2008.

Basavaraj S. Benni, Rawat: *Dairy Co-operative Management and Practice*, Delhi, 2005.

Baudrillard: *The Animals: Territory and Metamorphoses. Simulacra and Simulation.* Ann Arbor: University of Michigan Press, 1994.

Carroll, R. L.: *Vertebrate Paleontology and Evolution.* W. H. Freeman and Co., New York, 1988.

Clutton Brock Juliet : *Horse power: a history of the horse and donkey in human societies*, National history Museum publications, London 1992.

Davis, A. : *Let's Eat Right to Keep Fit,* New York, U.S.A: Harcourt Brace Jovanovich, Inc., 1970.

DeGrazia, David: *Animals Rights: A Very Short Introduction.* Oxford: Oxford University Press, 2002.

Ensminger, M.E. : *Dairy Cattle Science,* The Interstate Printers & Publishers, Inc., Danville, 1980.

Escobar, Roberto Calle: *Animal Breeding and Production of Camelids*, Lima, Peru, 1984.

Flowerdew, J. R. : *Animals: Their Reproductive Biology and Population Ecology*, Cambridge Univ. Pr., New York, 1987.

Godthelp, *Animal*: *Riversleigh. The Story of Animals Reproduction in Ancient Rainforests*, Reed Books, Balgowhah, 1991.

Greene, H. W.: *Mode of Reproduction in Lizards and Snakes of the Gomez Farias Region*, Tamaulipas, Mexico. Copeia, 1970.

Hacker, J.B. : *Nutritional Limits to Animal Production from Pasture,* Farnham Royal: CAB, 1981.

Hall S., Clutton Brock Juliet: *Two hundred years of British farm livestock,* Natural History Museum Publications, London 1988.

Jha, S N : *Dairy and Food Processing Plant Maintenance: Theory and Practice,* International Book Distributing, Delhi, 2006.

John, Prince: *Dairy Farming: Being the Theory, Practice, and Methods of Dairying*, New York, 1888.

Koli. P A: *Dairy Development in India: Challenges Before Co-Operatives*, Shruti Pub, Delhi, 2007.

Krieger, Maggie and Richard: *Secrets of the Andean Alpaca - The Field Guide*, Saltspring Island Llamas and Alpacas, 1994.

Law, Barry A.: *Microbiology and Biochemistry of Cheese and Fermented Milk*, London: Blackie Academic & Professional, 1997.

Mathialagan, P : *Textbook of Animal Husbandry and Livestock Extension,* International Book Distributing Co, Delhi, 2005.

Matthews, L. H.: *The Life of Animals Reproduction*, London, Weidenfield and Nicholson, 1969.

Nyholt, D.H. : *The Vitamin & Herb Guide,* Alberta, Canada: Global Health Ltd., 1992.

Patrick, L.: *An Introduction to Medicinal Chemistry*, Oxford University Press, Delhi, 2009.

Raymond, F., Redman, P., & Waltham, R. : *Forage conservation and Feeding.* Ipswich: Farming Press, 1986.

Rhykerd Charles L. : *The Cycles of Plant and Animal Nutrition*, Scientific American Books, San Francisco 1976.

Seidel, S.M.: *New Technologies in Animal Breeding*, Orlando, FL, Academic Press, 1981.

Shagufta Jamal and H P S Arya *: Participatory Rural Appraisal in Agriculture and Animal Husbandry : A Training Manual*, Concept, 2004.

Thompson, Paul B.: *Food Biotechnology in Ethical Perspective*, Aspen, CO: Aspen Publishers, 1997.

Thornhill, Nancy W.: *The Natural History of Inbreeding and Outbreeding*, Chicago Press, 1993.

Verma, S.R. : *Nature: Fish Genetics and Biodiversity Conservation*, Conservators, Delhi, 1998.

Vikram V. Mistry: *Cheese and FermentedMilk Foods*, Great Falls, Va.: Kosikowski, 1997.

White, M.J.D.: *Animal Cytology and Evolution*, Cambridge, Cambridge Univ. Press, 1954.

William Alec: *The Dairy Chemical Industry*, Longman Group Limited, London, 1971.

Yadav, Manju: *Mammalian Development*, Discovery Publishing House, Delhi, 2008.

Index

❑❑❑